# *Great Discoveries and Inventions by African-Americans*

**Volume One**

*by*

***David M. Foy***

*"People will never look forward to posterity who never look back to their ancestors . . ."*

*—Edmond Burke*

APU Publishing Group
P.O. Box 1137, Edgewood, MD 21040
Telephone: 410-538-7400

Library of Congress Card Cataloging Data: 97-69498

Foy, David M., 1936—
*Great Discoveries and Inventions by African-Americans*
1. African-American Inventors 2. Afro-American History 3. Great African-Americans 4. Inventions, in America 5. African-Americans, in business

ISBN: 1-878647-47-4

For additional information or to inquire about engagements or interviews with the author, David M. Foy, contact him at 919-510-5998, ext. 2222.

The illustrations of the inventors are by Reginald D'von Sharpe, an artist who lives in Raleigh, North Carolina. Mr. Sharpe can be contacted by writing to him at: 3948 Browning Road, #206, Raleigh, N.C. 27609.

Second Edition

*Dedicated to*
***Dr. Sonja Haynes Stone and***
***Dr. John Hope Franklin***

*. . . to* ***Dr. Sonja Haynes Stone,*** *in memory of her life's work and spirit as a dedicated professor of Afro-American studies at the University of North Carolina at Chapel Hill, North Carolina. It is through her voice and inspiration that this volume is completed.*

*. . . to* ***Dr. John Hope Franklin,*** *a distinguished professor and lecturer of history at Duke University, Durham, North Carolina, and a recent recipient of the Professor of the Century Award. Former Governor Terry Sanford of North Carolina made the presentation for Duke University, The University of North Carolina at Chapel Hill, North Carolina Central University in Durham and North Carolina State University in Raleigh. His work,* ***From Slavery to Freedom*** *has been used as a text book at North Carolina Agricultural and Technical College, now known as North Carolina A&T State University in Greensboro, North Carolina since 1955.*

*Dr. Franklin, African-Americans are grateful to you for your consistency and excellence in providing an accurate account of our history in print.*

# *Acknowledgments*

To my children, Shari, Lasalle and Jeffrey, whose growth is nurtured through an understanding of their "roots." To all of my ancestors, including paternal great grandparents, Robert Pearce (1878-1954) and Georgia Pearce (1882-1952).

Also to:
My paternal Great Great Great Grandfather,
Thomas Foy (1799-?).
Great Great Grandfather, Jacob Foy (1836-1915)
Great Grandfather, Miles Foy (1868-1937)
Grandfather, Reverend E.J. Foy, Sr. (1891-1955)
and my Father, E.J. Foy, Jr. (1913-1965)

And to:
The 1997 Senior Class of the James B. Dudley High School in Greensboro, North Carolina, and to Nancy Tice, Jennifer Calloway, Randy Williams, Linda Wright, all of Raleigh, North Carolina, who assisted me in the completion of this work.

# *Preface*

In this compelling volume one, *Great Discoveries and Inventions by African-Americans*, the facts are unobtrusively woven into a piece rich in the heritage, but little known knowledge, of many inventions and innovations of people of African ancestry.

Many of these inventions have never been promoted or accredited to Blacks. David M. Foy is the author of this definitive work and has drawn on extensive research and knowledge to produce a significant contribution of facts to a people often denied information about themselves.

Students, parents, educators, clergy, and the general reader of all ethnic backgrounds will have a better understanding and a clearer perspective of Black contributions through this informative book specifically designed to educate and enlighten.

***Grover Bailey***
Co-Publisher/Editor
*Dimensions News*
Raleigh, N.C.

# Contents

# *Foreword*

Although Black history is not an integral part of school curricula, our schools should acknowledge the substantial contributions by African-Americans to this nation, especially inventions which have been glaringly omitted.

Belatedly, many are learning more about Frederick Douglass, Booker T. Washington, Harriet Tubman, Sojourner Truth, Martin Luther King, Jr., Malcolm X and other civil rights leaders. However, neither children nor adults know very much about inventors whose ingenuity created such commonly used items as the ironing board (Sarah Boone, 1892), the mop (Thomas W. Stewart, 1893), and the clothes dryer (G.T. Sampson, 1892). Clearly, these, and numerous other practical discoveries, grew out of the hardships Black people endured in the workplace.

Promoting these "unsung heroes," David Foy of Raleigh, North Carolina, has a work-study guide: *Great Discoveries and Inventions by African-Americans,* that depicts some of the important creations of these noble personalities of genius and their contributions to humanity.

African-Americans played a major role in industrializing the United States by inventing tools and machinery to aid in a peoples exhaustive struggle to find ways to lighten their load. Lewis Latimer, a brilliant and versatile inventor, gave the world the apparatus for cooling and disinfecting. His device consisted of a fan and cooling coil used for purifying air and cooling hot rooms. It was the forerunner of the air conditioner. Latimer was also instrumental in developing the carbon filament for the light bulb. While working with Alexander Graham Bell, he wrote the first book on lighting systems, and managed the installation of the electric light in the state of New York. From Latimer's genius, such useful items as the first water closet for railroad cars, hat and coat racks, and the globe for electric lamps were created.

The list of Black inventors is extensive and should be a part of education in the public school system and other institutions of learning. *Great Discoveries and Inventions by African-Americans* is certainly an engaging work on an important subject.

***Andrew Young***
Former Ambassador to the United Nations,
Former Mayor of Atlanta, Georgia,
Civil Rights Leader, and Author

# Introduction

The African-American intellectual and creative presence in the development and invention of technical innovations, as well as the manufacturing of products, is not a recent phenomenon. This cultural heritage had its beginning in classical Kemet (Egypt), 4500 B.C. These African ancestors of the African-Americans were educated in the Kemetic (Egyptian) Secret Educational System. Their African Rites of Passage educational system included the study of writing, arithmetic, philosophy, geometry, astronomy, art, music, religion, engineering, medicine, and the secret sciences. They were, in fact, the designers and builders of the ancient pyramids, temples, libraries, obelisks, sphinxes, and monuments to the gods.

The African-American's Black ancestors in later years created and administered great empires in Africa. There were the vast empires of Ghana, Mali, Songhay, Abomy, Ashanti, Mossi, and Yaruba in West Africa. There were the kingdoms of Kuba, with Shyaam the Great, and Angola with Queen Nzinger, in Central Africa. Shyaam was an economic revolutionary who transformed his nation into perceiving the value of economic development. In East Africa, there was the Great Zimbabwe, and the Great Zulu Empire in Southern Africa.

It should not be surprising, then, that the great African-American individuals whom David Foy has presented in this important book, have produced products and inventions that have made modern life more enjoyable. They came from great African people.

There are important reasons, however, that these African-Americans are not as well-known and famous as the great White American heroes. These reasons are economic, religious, cultural, and educational in nature. Africans were first brought to the Americas and enslaved for purely economic reasons. They should be properly referred to as "economic prisoners," and not as slaves.

The only way African-Americans could be continually used to make Whites wealthy and powerful, was to make them feel inferior and not as intelligent as Whites: it was the ongoing process of the Psychotechnology of Brainwashing. This brainwashing was done to make them forget and to not even desire to know anything about their African heritage in economics, religion, culture, and education.

African-Americans were, and still are, deliberately miseducated and under-educated in the educational systems, as well as in the general society. Whites and Blacks are guilty of these acts. However, most African-Americans are not even aware of their complicity.

A society that wishes to achieve greatness, respect, and independence must at all times know its own culture, religion, education, rituals, and heroes. If this is not observed, "tales of lions will always glorify the White hunters."

This is a refreshing book that is not only educational, but motivational and spiritual. The spirits of the individuals found within this book still live on within our youth and adults. Like seeds buried beneath the ground, they only need water and sunlight to burst into physical and spiritual growth.

David Foy has offered you the spiritual water and sunlight of some of our African-American heroes. Now use them well, drink deep, and lift up your arms to your own growths and achievements. Sankofa.

NANA KWABENA F. ASHANTI, Ph. D.
Counseling Center and Visiting Professor of African-American Studies, North Carolina State University, Raleigh, North Carolina

# Chapter 1

# *Joe L. Dudley*
## *(1937-Present)*

## *Inventor of 200 Ethnic Hair Care Products*

*"I think the keys to success are desire and persistence. Consistent to be persistent. Success is a journey, not a destination..."*

*—Joe L. Dudley*

When entering the state of North Carolina from its far western tip, one experiences the majestic splendor of its lofty mountains, and moving easterly to its other end, one can see and feel the flow and ebb of the tides of the Atlantic ocean.

Mr. Joe L. Dudley, a native North Carolinian, was born in the Flat Lands between two extremes: mountain peaks and ocean front of Aurora, North Carolina, on May 9, 1937. He was born during the era when professional boxing was very popular with Americans. On September 24, 1935, Joe Louis Barrow, an African-American affectionately called the "Brown Bomber," won the world heavyweight boxing title. "Joe Louis" was highly esteemed as a respected boxer and role model by Black and White America, and Mr. Dudley's parents named their son Joe L. after this famed boxing champion.

One of eleven children, Mr. Dudley, was raised in a three-room farm house in the rural area of Halifax County. He suffered a severe speech impediment, failed the first grade and was labeled mentally retarded by his teachers. Nevertheless, Dudley's mother never gave up on her son. His early life revolved around his father's and grandfather's farm.

"All of us worked on the farms; we made our living from the earth," Joe Dudley recalls. "Life on the farm was isolated. We didn't have a radio, television or newspaper. Only after reaching age fifteen or sixteen, on Saturdays and Sundays, could we go to town."

When his grandfather moved from his farm into Mr. Dudley's parent's home, the family unit consisting of three adults and eleven children, all shared that three-room house. This created an extremely overcrowded living condition. The small farm house had no running water, only an outside toilet, and no screens at the doors or windows. Their bath water was brought in from the well and heated. The girls in the family would bathe first and the boys would bathe in the same water once the girls had finished. Insects, flies and mosquitoes invaded the home, making living conditions even more rustic. At night, during the summer or winter, the sky, stars or snow flakes, were visible from Dudley's bed.

When Dudley was a young boy, the family's house burned to the ground. After the loss of the house, his parents urged him and his brothers and sisters to seek an education: "something that could not be destroyed." His father, despite only having a fifth grade education, set an example for the children by reading.

Growing up in North Carolina, Dudley says he struggled through school and was held back a second time before reaching the eleventh grade. However, his mother, Clara, strongly supported Joe and told him, "When slow people get it, they've got it." He made it through high school and later attended college. During his educational years, Joe also discovered the value of self-motivated learning. While in the eleventh grade, he decided that he would help himself and began to study materials he had already studied from the first through the tenth grades. At that time, he concluded that anything he wanted to learn could be found in the library.

Beginning at square one, Dudley began re-reading his elementary school books. "I started with the first grade, continued to the second grade, and went all the way up to the tenth grade," he says. While he didn't excel academically at school, he discovered and attained self-esteem.

Between Joe Dudley's interactions in the public school and library arena, the conjectured message of him being slow and retarded spread very quickly. When this innuendo reached the young lady in whom he had interest as a special friend, she immediately told him that "I want beautiful and intelligent children and not those who would be ugly and dumb." This statement on the part of the person in whom he had confidence and interest served as a catalyst for him to reach deeper into himself to find the resources—emotional and spiritual—to become more self-determined and positive. This important event in the life of this youthful student, Joe Dudley, became the turning point in his life.

Dudley stated, "I saw a brighter future. I knew I could make something of myself." After high school, Dudley enrolled at North Carolina A&T State University in Greensboro, North Carolina. While his older brothers were pursuing their degrees in veterinary medicine and engineering, Joe felt more comfortable doing what he was familiar with, farming; thus, he majored in the science of poultry. To help pay his tuition, he collected eggs and fed the chickens on the University farm. He also did domestic chores for a professor on weekends.

One summer, while living with an aunt in New York, during his struggles to find employment, Joe met a door-to-door salesman selling Fuller Products, which was a business enterprise owned by a Black entrepreneurial "brainchild." Dudley invested ten dollars for a sample kit and began selling products door-to-door. "I was very shy because of my speech impediment," states Dudley. "But I felt this was doing something where I could be my own boss and earn money while learning."

Dudley continued to sell Fuller Products door-to-door after he returned to college and changed his major from poultry science to business administration. Through meeting other salespersons working for Fuller Products, he sharpened his sales skills and increased his sense of self-worth. He said, "They were a group of people who constantly set a good example and motivated me."

In 1960 he met Eunice Mosley, also a salesperson for Fuller Products. A year later, they were married. Following college graduation, Joe and Eunice moved to New York City where they planned to live permanently. For five more years, Dudley worked for Fuller Products, where he met the legendary S.B. Fuller of Chicago, Illinois, who became his friend and mentor. Fuller, the inventor and supplier of ethnic hair-care products began experiencing inventory problems with his company. Dudley, becoming discouraged, decided to return to North Carolina to raise hogs. Fuller, however, convinced Dudley to stick with it. "Raise people instead," instructed Fuller, "people talk back."

In 1967, Dudley and his wife moved to Greensboro, North Carolina, and opened their very own Fuller Products distributorship. When the parent company in Chicago was unable to meet the increased demand for the products, Dudley and his wife began making their own products! They worked during the night packaging them in old mayonnaise jars and empty containers collected from beauty operators. By 1976, they opened a chain of beauty supply stores, beauty salons and a beauty college. They opened fourteen stores and employed more than four hundred people.

At the request of founder, S.B. Fuller, in 1976, the Dudley's moved to Chicago to help Fuller in the management of his company. They continued to operate Dudley products out of Chicago and in 1980 Fuller Products and Dudley Products consolidated under Dudley's name. A decision was made to close the beauty supply stores and shift the company's focus to manufacturing. This decision resulted in improved control, distribution, and quality of the products. It also enabled them to build a national company.

In 1984, they moved the company back to Greensboro. As of the year 1996, the firm manufactures two hundred products in an 80,000 square foot facility that serves as headquarters, and sells these products to 30,000 cosmetologists.

Commitment to his employees is of paramount concern to Dudley. The company's monthly sales meetings focus not only on product knowledge and selling techniques, but also on personal development principles ranging from building self-confidence to the importance of looking within oneself to realize who you are and what you can become. Dudley further encourages his employees and students to believe that "The keys to success are desire, persistence, and consistency . . . success is a journey, not a destination."

Today, there are Dudley Cosmetology Universities in two key North Carolina cities, Greensboro and Charlotte; Chicago, Illinois; and Washington, D.C. A ten month program is designed for beginning cosmetologists and advanced courses for experienced cosmetologists are also offered at the Kernersville, North Carolina, location. The company, in addition, operates a travel agency, a convention center, a publishing division, and a hotel at the Kernersville location. "We are on a mission to do more than just sell products. In the future the need for haircare education will increase worldwide," says Eunice Dudley.

Joe Dudley's greater concern is to leave a legacy in his industry, rather than to just experience things like his annual practice of setting sales and growth goals for the company. "I want to empower people in this industry . . . and show them that if you work hard enough and believe in yourself, you can make it . . . "

Mr. Dudley is recognized nationally as a successful entrepreneur, lecturer and humanitarian. One of his humanitarian efforts is the Dudley Fellows Program which provides incentives and mentors for local Black high school students. Graduating participants of the Program are awarded full scholarships to North Carolina A&T University or Bennett College. This Program was recognized by President George Bush as his 467th Point of Light.

Dudley's most cherished award is the Horatio Alger Award, whose recipients are honored for their commitment to helping young people, and encouraging hard work in the face of adversity.

Only in America! Only in America—could a common citizen make a ten dollar investment and gross more than 100 million dollars in a few years!

# Chapter 2

# *Lewis H. Latimer*
## *(1848-1928)*
## *Inventor of the Carbon Filament for the Light Bulb*

*"I was one of the pioneers of the electric lighting industry from its creation until it had become worldwide in its influence . . ."*
—Lewis H. Latimer

The invention of the electric light bulb is indeed a creation that placed the Western world and the global society in a forward technological mode. Mr. Lewis H. Latimer, an African-American, worked with Thomas A. Edison to invent the first practical light bulb that is used today to light homes, businesses, streets, buildings—inside and outside, etc. Latimer also drew the plans for Alexander Graham Bell's first telephone patent.

Thomas Edison used a team of scientists to work out the problems in the invention of the light bulb, and Lewis Latimer was an important member of that group. The laboratories where much of the scientific research for this project was done were located in Menlo Park, California, and West Orange, New Jersey. These centers became models for many research centers, which were integral parts of life in the twentieth century. This team used an approach in scientific discovery and technology that is still widely used today. Lewis Latimer was the primary researcher that made Edison's lighting system possible. The creation of certain inner parts of the incandescent light bulb are directly attributed to this African-American scientist.

In December of 1928, Mr. Latimer died and the Edison Pioneers released to the press facts about his life and contributions in the field of electric lighting. The press release read as follows:

> *Mr. Latimer successfully produced a method of making carbon filaments in the maxim electric incandescent lamp which he patented . . . He was of the colored race, the only one in our organization, and was one of those to respond to the initial call that led to the formation of the Edison Pioneers on January 24, 1918.*

The 1800's was an era in which many Black slaves left the South and traveled North to escape to freedom. The father of Lewis Latimer, George Latimer, was one such slave. George lived in Virginia, but escaped North. After his courageous flight, he started a new life in Boston, Massachusetts. Here, George Latimer met and married another fugitive slave from Virginia. This marriage produced four children, a girl and three boys. Lewis Latimer was born in September, 1848, the youngest of his family. His boyhood did not lend itself to the belief that someday he would be famous, especially being the son of former slaves.

While Lewis attended elementary school in Boston, he also worked several hours a week in his father's barbershop, and as a boy was very creative. He learned to hang wall paper and became effective at following his father's direction. Early in life, around age ten, he dropped out of school to work full-time for his father. George Latimer, for unexplainable reasons, left his home, abandoning his wife and children. Mrs. Latimer was unable to support the children and sent Lewis and his brother to a farm school in the western part of Massachusetts.

For several years Lewis remained at the farm school, until one day, his brother William came by to visit. His brother suggested that they run away to return to their native Boston. Lewis did not enjoy the chores of the farm so he readily agreed to leaving the school. Lewis shared with his brother that he wanted to find a job where he could use his mind instead of his muscles. The distance from the school back to Boston was roughly eighty miles. Lewis and William carefully planned the escape from the farm school hoping not to get caught by school authorities who perhaps would be searching for them. They stole rides on the railway, ran, walked, begged for food and water and arrived in Boston within a few days. This running away experience by Latimer could have given him a good feel of how his father and mother felt as they escaped from slavery in Virginia, going to Massachusetts to their freedom. Latimer was only thirteen years old when he arrived back in Boston. His work for a while included working in the homes of wealthy people, waiting tables and also as an utility carrier in a law firm.

When the Civil War started in the early 1860's, Lewis was fifteen years old. By this time in his life he had grown tired of odd jobs and enlisted in the United States Navy. This was done through telling an untruth regarding his age. In the Navy, he served as a cabin boy. When the war ended in 1865, he returned to Boston. It was difficult for him to find employment for a period of time. He finally secured a job in the office of patent solicitors, Crosby and Gould. He was paid three dollars per week, overseeing drafting tables where inventors made detailed drawings of their inventions. Lewis Latimer was fascinated by the work of the draftsmen. As he watched them carefully make drawings for their inventions, he also observed the books and other materials being used. He started to purchase second hand books that gave him elementary instructions on drawing. After saving enough money, he bought the tools and instruments to do his own drawings. It was through observing the draftsmen that he learned to use the instruments.

At night he would use his time practicing what he had observed the draftsmen do while working. He knew that "practice makes perfect," so it was not unusual for him to practice drawing into the late hours of the night. After a period of time, Lewis' boss saw him at the drawing board and was so impressed with his work that he gave him the position of junior draftsmen. This was a promotion from his office assistant position. Shortly thereafter, he was working eight hours a day and, on some occasions, made working models of inventions to go along with the drawings that were necessary to go to the patent office. Latimer worked for this particular drafting company for eleven years earning eighty dollars a week.

It was in 1876, while Latimer worked at Crosby and Gould, that Alexander Graham Bell commissioned them to make a drawing of his telephone system patent. Latimer was assigned to draw Bell's telephone system. Mr. Bell had written very specific instructions on how he wanted the drawing of the telephone done, and Latimer followed the directions precisely.

When Latimer had reached the age of thirty, he had grown restless regarding his accomplishments as he had a very creative and active mind. He had also become a very skilled patent draftsmen. Lewis Latimer's influx of thoughts centered for the most part around improving upon the work he was doing and becoming an even greater inventor.

Some fifteen years later, Latimer moved from Boston to Bridgeport, Connecticut, to live with his married sister. His first work opportunity there was in a machine shop. And his first assignment was once more mechanical drawings. An inventor named Hiram Matim must have been shocked when he walked into a mechanical shop to find a man of color working as a draftsmen. However, reviewing Latimer's work, he chose Latimer to do drawings for him. Mr. Matim was the inventor of the machine gun. He also was the chief engineer of the United States Electrical Company and quickly realized that Mr. Latimer was a talented draftsmen.

Mr. Latimer worked for Matim as a mechanical draftsmen and learned everything he could about the electrical light construction and operation. When Thomas Edison invented an electric light, Latimer immediately started to experiment with ways to perfect and improve Edison's creation. Mr. Latimer left the New England area in 1880, moving to New York with Mr. Maxim, who relocated his company.

After his move to New York, Maxim assigned Latimer the challenge of producing the carbon filaments for electric lamps. The material in the electric light bulb which causes it to glow is the filament. The glowing process occurs when an electric current runs through it. All light bulbs (incandescent lamps) for over a quarter of a century had carbon filaments.

In an incandescent lamp, the electric current heats up to almost white heat. The heat gives off the light we see. The amount of light the bulb gives off is determined by how hot the filament is heated. The life span of a light bulb is also determined by how hot the filament is heated. In the opinion of Thomas Edison, 600 hours of use of a filament was a good average. Lewis Latimer felt this was not good enough.

Mr. Latimer continued to work diligently on the quality of the filament material. He eventually was able to raise the operating temperature of the bulb without loss to the life of the filament. In 1882, Latimer received a patent on his electric light filament. His patent was received after long hours of improving the quality of the filament. He arrived at a method of making an improved and less expensive carbon filament for Maxim's electric lamp. With this improvement, Latimer made it possible for more people to purchase electric lamps and to have electric lighting.

After Latimer had received a patent for an improved

filament, he was appointed as the Chief Electrical Engineer for Maxim's company. He soon was assisting in installing incandescent electric light plants in major cities of the United States and Canada. He also supervised the production of his own carbon filaments which were used in lights in railway stations in Montreal, Canada and New York.

Mr. Maxim later requested that Mr. Latimer go to London, England, to establish an incandescent lamp factory. There, Latimer gave directions to workmen in every aspect of making light bulbs and blowing glass. Within a short time span, Mr. Latimer returned to the United States where he designed and manufactured a lamp that bore his name. The lamp can be located in a collection of incandescent lamps at the Henry Ford Museum in Dearborn, Michigan. One of 800 lamps in the collection is Latimer's. This collection of lamps is an example of the development of electric lighting systems. Latimer accepted another opportunity at this point in his life to work for Edison as a senior draftsmen. For thirty years he remained with Edison's company. Mr. Lewis Latimer continued in his adult life to have a great deal of admiration for Thomas Edison and his work. And he wanted everyone to know as much about the lighting system as possible. In 1894, Latimer wrote his first book on the electric lighting system. This book was entitled, *Incandescent Electric Lighting*, and was a usable description of Edison's system.

In his book, Latimer specifically described how light is produced by heating a filament to incandescence and how various parts of the electric light work. He described that if electric current can be forced through a substance that is a poor conductor, it will create a degree of heat in that substance. The intensity of the heat is determined by the quality of the electricity forced through it.

The chemical properties of copper and platinum wires readily conduct an electrical current, the carbon filament offers a great deal of resistance to its passage and for this reason becomes very hot, in fact it is raised to white heat or incandescence, which gives its name to the lamp. There is apparently doubt or wonder why this thread of charcoal is not immediately consumed when it is in this state, but this is accounted for when one remembers that "without oxygen of the air, there can be no combustion and every trace of air has been removed from the bulb which is so thoroughly sealed as to prevent admission of air about it."

The incandescent bulb does not last forever, for the reason that the action of the electrical current upon the carbon has a tendency to divide its particles and transfer them from one point to another so that the filament gives way at some point. Thus, most of the lamps last for 1,000 hours, which amounts to between four to six months of use.

Mr. Lewis Latimer was more than just an inventor

who improved the quality of life for people who would use his incandescent light bulb. Yes—streets, office buildings, railways, subway systems, homes, etc., are being provided light, and living for all segments of society is more comfortable because of his contributions! Beyond these inventions, his work in civil rights organizations also helped to make living more equitable for all. He also taught the English language to immigrants in New York City.

Near the end of his life, as Latimer was growing much older, he expressed himself in other ways: writing poetry, music and painting. The theme of love ran through much of his artistic undertakings. He believed, "Love is to the human heart, what sunshine is to a flower, and friendship is the fairest thing in this great world of ours."

Only in America could an African-American, in one generation, improve from slavery and make such notable contributions to his specific society and to the world at large.

# Chapter 3

# Lewis Temple (1800 - 1854)

## Inventor of the Harpoon Toggle Iron

*"Studying the past defines the future . . ."*

—Unknown

The Atlantic Ocean, with its sprawling shores extending throughout the natural borders from Connecticut to Maine, placed this region in a natural position for the whaling business and its economic contributions. Since the formation of the Thirteen Original Colonies of

America, the New England States have been engaged in the fishing and whaling industry.

Mr. Lewis Temple was believed to be the man to revolutionize the industry during those early years of the young colonies. The whaling business, it is thought, was brought to America by early European settlers. The year of 1746 is used as the date to denote its arrival here. And very quickly New Bedford, Massachusetts, became known as the whaling capital of the region.

Mr. Lewis Temple, an African-American, became the man to invent a tool called the Toggle Iron, which made the catching of whales more certain. Before an improvement of the harpoon by Temple's thoughts and hands, whales would often escape after having been hooked. Temple's invention gave the harpoon a movable head, which functioned in such a way as to prevent a harpooned whale from slipping loose from the hook and swimming back into the water. Temple's harpoon operated in such a way that it "locked" into the whales flesh, and the only way to free it was to cut it from the whale after it had been killed. The whaling industry had been around one hundred years before this invention. The new instrument made catching whales more sure. In 1848, this new invention became a reality. Before this era, all of the products which came from whales—oil, meat and whale bones—were scarce since so many of them escaped back into the waters of the ocean.

Lewis Temple was born in Richmond, Virginia, in 1800, a state in which the holding of slaves was legal. Whether Temple was a slave or free man is uncertain. His movement from Virginia to Massachusetts might indicate that he was free to move North. It was around 1830 that Lewis Temple ended up in New Bedford, Massachusetts. This city was one of the northern places where slaves who were runaways ended up.

The underground railroad was a system whereby slaves in the South could escape by going North, and finally into Canada, to freedom. During this period, persons who were opposed to slavery used their homes and farms to hide slaves until they could escape further North and in many instances, ended up in Canada.

As a special note, in the early 1800's in Wilmington, North Carolina, there was in existence an underground railroad station at the Gregory United Church of Christ. This station had existed before the riots in the mid 1800's in that city where Black and White lives were lost and property damage was extensive.

The peak of the underground railroad system was between 1810 and 1850. It is estimated that over 100,000 men and women escaped to their freedom during this period. The records are not very clear, but it is thought that Temple ended up in the New England states during this time. It is also estimated that there were 315,000 free slave people in America. Temple used his freedom to begin the process of becoming a successful African-American in the United States.

In the year of 1836, Lewis Temple was located in the small fishing town of New Bedford, Massachusetts. He was married and his family included three children. Somewhere in his background he learned the trade of being a blacksmith. Although the record is not clear on how he learned this trade, it is known that many Black men who were living in America from the continent of Africa had brought various trades and skills to their new world. In 1845, Temple had developed a blacksmith shop and was doing well in his business. Since Temple obviously had innate ability and skill for creating things, he apparently used his shop to invent items for the fishing industry, which was the third largest in the New England area.

It is more than likely that in his shop Temple met and talked with fishermen and sailors who came from the seas, resulting in the development and shaping of his thinking methods for improving the tool to catch whales. Out of his talks with people from the sea waters, he finally came up with the idea to create a new type of harpoon.

Temple, with his innovative mind, carved out a way to produce a device so that once a whale was caught, the hook would not pull from the flesh allowing it to escape back into the waters. The toggle harpoon he developed worked in such a way that it secured the whale after being caught. Up to this time in the whaling industry in America, the harpoons then in use had been passed down by the English and Dutch in the whaling business. There were two basic types of harpoons used in the industry. Single flue and double flue harpoons, both came from the fishing business in the old world. (If one would take a present day fishing hook and examine the end of it, there would be found an example of a flue.)

The Lewis Temple Toggle Iron was movable and not fixed into place. In the harpoons that had come from England the head of the iron was fixed and immovable. The basic change that Lewis made in his Toggle Iron, was its mobility. He mounted it on a shank so that it revolved when it entered the flesh of the whale. It toggled, turning at a right angle, locking itself in the whale's body. With this method, when the whale jerked and pulled away to return to the water, it could not escape.

For a brief period after Temple's invention, fishermen were slow to take advantage of this new technology. After a brief span of time, however, the use of Temple's iron caught on and proved to be an invaluable tool in the whaling business. In the end, Temple's iron was established and known to be superior to previous whaling instruments.

Several other fishermen prior to Temple had also attempted to create a tool that would be more effective in the whale fishing industry. However, none of those tools were as advanced as Temple's new creation. Temple's invention of the toggle harpoon with a double flue is what made whale fishing in America more productive. Had his invention not become a reality, whalers would have had to continue to gamble on the number of whales caught during a given fishing trip.

People in America with blood of African lineage were unable to have their inventions patented before the addition of the Thirteenth and Fourteenth Amendments to the U.S. Constitution. It was after these additions to the Constitution that the patents of their creations became a realization. For reasons that are unknown, Temple never had his invention patented. It is known that other blacksmiths in the New Bedford area copied Temple's Toggle Iron. It is also established from the whaling industry records that one blacksmith in the area made some 30,000 toggle harpoons over a twenty year period.

In the early to mid-1850's, Temple was making a good living in his blacksmith shop manufacturing his Toggle Iron, but had he patented the invention, his monetary gains would have been much greater. Despite the odds against Temple, he continued to be motivated, working hard to the extent that he started to build a larger blacksmith shop next to his existing home.

In the process of having his new shop built, which was being made from bricks, an unfortunate thing happened. In the darkness of night, Temple was walking around in the area of the new construction—making an awkward step he fell and injured himself. He was injured so seriously that he could no longer work at his skill of making toggle irons. He never recovered from his injury, dying in the mid-1850's at the age of fifty-four.

At Temple's death, his estate was evaluated and his wife and children were left in a real financial hole. The new blacksmith shop which he had started to build was never completed. After selling his personal property which included his tools and other equipment, his home, the incomplete building and paying off of his debts, nothing was left over for his family to exist on.

The positive side of Temple's life and death is that he left the broader society a powerful and ongoing legacy. The fishing and whaling industry was left his progressive and effectively made toggle iron. And beyond the fishing industry, the whole human family in America in some small way benefits from the genius of his creative mind.

# Chapter 4

# *Granville T. Woods*
## *(1856-1910)*

## *Inventor of the Telephone Transmitter*

*"Education sows not seeds in you, but makes your seed grow . . ."*

—Kahlil Gibran

Mr. Granville T. Woods, a man of African lineage in America, was born in Columbus, Ohio, in 1856. His birth date was eight years before the signing of the Emancipation Proclamation by President Abraham Lincoln. This document is the instrument which legally outlawed slavery in America. Up until this time in U.S.

history, the constitution had made an allowance for Black people to be treated differently from other racial groups.

Even in the first census tabulation in 1790, Blacks were counted as two thirds of a human being, as numbers in the households of the slave owner. Mr. Woods' birth during this era obviously indicated that he would have great odds to overcome to be successful in America. His efforts to become superior to the external conditions in America started with his energy to get an education. His elementary schooling began in Columbus, Ohio. He attended school until he was ten years old, dropping out at an early age to work in a mechanic's shop.

It was during his first work experience that he started to develop a long and keen interest in the field of railroads, trains and other related equipment. Woods, very early in his boyhood, began to show an astute interest and ability towards learning. He was even anxious to learn and appeared to have had an aptitude for grasping materials in the areas of electrical and mechanical engineering. He was an avid reader with a special aptitude in the physical sciences.

With what appeared to be an aptitude for the mechanical and electrical sciences, he transferred what he learned in one job experience to another. During his youth, Woods changed jobs frequently. In addition to his strong interest in reading, he would also pay shop foremen on the various jobs he held to instruct him on the details and specifics of their work. When Woods was sixteen years of age, he left Ohio traveling West. For a period of time, he had difficulty finding work while in the state of Missouri. He finally got a job as a fireman with a railroad company. The fireman's job on the train was to keep the furnace stocked in order for the train to receive energy to move. He worked for the Iron Mountain Railroad in Missouri. His job there ended up in him being promoted to an engineer.

Granville Woods continued by using much of his spare time reading books on electrical configurations and designs. He moved within a short time span to the state of Illinois, securing employment in the city of Springfield. The new work environment allowed him to learn the art of molding and shaping steel and iron. While in his early twenties, Woods left Illinois ending up in the East at a technical school. In this school setting for two years, he continued to increase his knowledge in mechanical and electrical engineering. Though his training helped to refine his skills, he still worked long hours while attending school. His classes of study were in the evenings while he spent time during the day in the machine shop.

After Granville had completed his time at the two year technical school, he then moved on to take a job at sea as an engineer aboard the Ironside, a British steamer. This job carried him to almost every continent in the world. He moved from his position at sea after two years.

Returning back to his interest in railroads, he secured a job in Cincinnati, Ohio, handling locomotive engines.

During Mr. Woods years of study and travel he was often rejected and scorned because he was a man of color. At times he was depressed and stressed out as a result of this, but he never allowed this negativism to control him. He believed with his training and aptitude for electricity and engineering, that racial discrimination could not hold him back.

Often, on the jobs where he had been employed, he did not advance because of his color. He became increasingly aware of this type of spirit in America and opened his own business and continued to work hard to develop the inventions he envisioned. He was still a young adult when he decided to open this business. He organized the Woods Electric Company in Cincinnati, Ohio. By 1884 his business took many of his initial patents. From this time, his inventions started to increase in number and in their value.

In the electrical/engineering worlds, Woods drew the attention of several of the large corporations in America: Westinghouse, General Electric and American Bell Telephone. His reputation grew to the extent that these corporations started purchasing information and devices from him.

The Bell Telephone Company purchased an instrument for transmitting messages by electricity. Mr. Woods describes the invention:

> . . . the operator uses a 'key finger' to irregularly make and break the circuit or to vary the tension of the electric current traversing the 'line-wire,' the key being operated by the varying pressure of the operator's finger . . . the message thus transmitted is received by an instrument known as a 'receiver' or sounder, which causes audible atmospheric vibrations in response to the pulsation's of the electric current traversing in the line-wire . . . my system (called by me 'telegraphony') entirely overcomes the failings of the ordinary key and sounder and has a wide range of usefulness. It was capable of use by inexperienced persons. For example, the operator cannot read or write the Morse signals, he has only (by means of a suitable switch) to 'cut' the battery out of the main-line circuit and 'cut' into a local circuit and then speak near the key. This having been done, the sounder of the receiving station will cause the air to vibrate in unison with the electric pulsations that will traverse the line-wire. The person at the receiving station will thus receive the message as articulate speech.

Mr. Woods discovered and invented many instruments, gadgets and devices which continue to serve cities and communities around America in so many ways. Other inventions will be noted later in this narrative.

Today, travel by train is safer because of the invention of a train telegraphic system. This invention, perfected by Woods to make traveling by train less of a risk, was called the Induction Telegraph System. This followed the principle of electrical induction. In the electrical induction process, there are two wires laid side by side—close but not touching. One wire is the primary (a) wire and the other, the secondary (b) wire. The length of both wires are the same. Wire (a) is conducting electricity in a parallel position beside wire (b). The transfer of an electric current between the two wires from primary (a) to secondary (b) is known as the principle of electrical induction. This method of transferring electrical current induced in one wire and floated into the other is the basic theory of the train telegraphic system. Granville did not discover induction telegraphy, but he transferred the general principal in the system to what he developed for railroads.

The number of train accidents were reduced once his new invention in telegraphy had been perfected. Mr. Woods described the invention in his patent application. He wrote about his induction telegraphic system:

> . . . it relates to systems of electric communications between two moving railroad trains or vehicles or between the same and a fixed station or

> stations, and transmits the signals to and from the vehicle by means of induction, whereby an electric impulse upon the line-conductor is caused to produce a corresponding impulse upon the similarly arranged conductor carried by the vehicle in close proximity to the line conductor . . . Any code of signals may be employed in my system of communication. A telephone receiver and a telegraph relay may be arranged so that either one may cut into the circuit when a signal is to be received.

Granville T. Woods is pointed out in Mr. Robert C. Hayden's work, *African-American Inventors,* as the inventor of a device for regulating electric motors. Many household gadgets and machines such as vacuum cleaners, washing machines, dryers and food mixers are run by electric motors. In an electric motor, electrical energy is changed into energy of motion or mechanical energy. In certain household appliances and gadgets with electrical power, it is necessary at times to change the speed of the rotating shaft of a motor without changing or deregulating the electrical voltage of its source.

Mr. Wood's invention of a device called a dynamotor was able to have the speed changed at the shaft of certain household appliances by adding coils of electrically resistant wires to the motors. These coils, known as resistances, would drain off some of the electrical power coming into the motor. In the instance of an electric bread toaster, the coils got hot very quickly and using them was a waste of electrical energy.

Between 1895 and 1900, Woods had his motor regulator challenged in the U.S. Patent Office to determine the true inventor. Only one other inventor in the United States had partially completed his creation.

Mr. Woods lived from 1856 until 1910 and his patenting record of different inventions surpassed all other persons, including those from different ethnic groups. In 1917, nearly ten years after the death of Granville T. Woods, the *Journal of Negro History* in an article entitled "The Negro in the Field of Invention" by Henry E. Baker, an examiner in the U.S. Patent Office, wrote the following: *"There is no inventor of the colored race whose creative genius has covered quite so wide a field as that of Granville T. Woods, nor one whose achievements have attracted more universal attention and favorable comment from technical and scientific journals both in this country and abroad."*

*"His name will be handed down to coming generations as one of the greatest inventors of his time . . . "*

—*Catholic Tribune, Cincinnati, Ohio, January, 1886*

## Other Great Inventions and Discoveries by Granville T. Woods

| Invention | Patent No. | Patent Date |
|---|---|---|
| Steam Boiler Furnace | 299,894 | June 03, 1884 |
| Telephone Transmitter | 308,876 | Dec. 02, 1884 |
| Apparatus/Transmission Messages by Electricity | 315,368 | Apr. 07, 1885 |
| Relay Instrument | 364,619 | June 07, 1887 |
| Polarized Relay | 366,192 | July 05, 1887 |
| Electro-Mechanical Brake | 371,655 | Oct. 11, 1887 |
| Railway Telegraphy | 373,383 | Nov. 15, 1887 |
| Induction Telegraph System | 373,915 | Nov. 29, 1887 |
| Overhead Conducting System/Electric Railway | 383,844 | May 29, 1888 |
| Electro-Motive Railway System | 385,034 | June 26, 1888 |
| Tunnel Construction for Electric Railway | 386,282 | July 17, 1888 |
| Galvanic Battery | 387,839 | Aug. 14, 1888 |
| Railway Telegraphy | 388,803 | Aug. 28, 1888 |
| Automatic Safety Cut-Out for Electric Circuits | 395,533 | Jan. 01, 1889 |
| Electric Railway System | 463,020 | Nov. 10, 1892 |
| Electric Railway Conduit | 509,065 | Nov. 21, 1893 |
| System of Electrical Distribution | 569,443 | Oct. 13, 1896 |
| Amusement Apparatus | 639,692 | Dec. 19, 1899 |
| Electric Railway | 667,110 | Jan. 29, 1901 |

| | | |
|---|---|---|
| Electric Railway System | 678,086 | July 09, 1901 |
| Regulate/Control Electrical Translating Devices | 681,768 | Sept. 03, 1901 |
| Electric Railway | 687,098 | Nov. 19, 1901 |
| Automatic Air Brake | 701,981 | June 10, 1902 |
| Electric Railway System | 718,183 | Jan. 13, 1903 |
| Electric Railway | 729,481 | May 26, 1903 |

# Chapter 5

# Garrett A. Morgan (1877-1963)

## Inventor of the Traffic Light System

*"People will never look forward to posterity who never look back to their ancestors . . ."*

—Edmond Burke

This man of color, and of African lineage in America, was born on March 4, 1877, in Paris, Kentucky. His birth date was during the era of the election of the eighteenth president of the United States, Ulysses S. Grant.

President Grant's administration was highlighted by

the development of the Transcontinental Railroad, the Amnesty Act, and the death of General George Armstrong Custer. This period in United States history immediately followed the signing of the Emancipation Proclamation by the eleventh president of the United States. President Abraham Lincoln had used his influence to end the institution of slavery in America. Mr. Morgan's mother had been freed following the signing of this document by the president in 1863.

Garrett A. Morgan grew up on a farm in Kentucky with brothers and sisters and attended school through the sixth grade. At the young age of fourteen, he left home traveling to the next state of Ohio. It was in the city of Cincinnati that he found his first job. For four years he worked as a handyman for a wealthy landowner. Morgan was anxious to learn and hired a tutor to help him with the English language. Four years later he moved to Cleveland, Ohio, remaining there for the rest of his adult life.

Morgan worked for a while in Cleveland as a mechanic adjuster for a manufacturer that made clothing. His enjoyment was tinkering with machines and he developed the skill in making repairs to them. The skill of repairing machinery afforded him jobs at various times. Several years after the turn of the century, Mr. Morgan opened his own business. His business provided the service of repairing and selling sewing machines. He was a very thoughtful man who quickly—within a two year period—added a tailoring component to his shop. He made several devices that could be added to the sewing machine, which allowed him to make suits, coats and dresses. His clientele included both men and women. He employed a total of thirty-two people for the manufacturing of these items.

Mr. G.A. Morgan was a multi-talented man. One evening before dinner, he was experimenting with a liquid that would give a high gloss to sewing machine needles. He also needed to discover something that would prevent the sewing machine needles from scorching cloth as it stitched. Some of the liquid substance being experimented with accidentally fell on a piece of pony-fur lying on his work bench. This incident lead him to ask his next door neighbor to let him try some of the liquid needle polish on his dog who had fur. The fur of his neighbor's dog became so straight it could hardly be recognized as the same dog. Mr. Morgan finally used some of the same liquid on his own hair and got a similar result. He continued to experiment with the needle polish until it was made into a creamy substance. He named it his "magic" cream for hair. Here, he had discovered the method of straightening hair.

He also organized another business, naming it the G.T. Morgan Refining Company. He started to market his "magic" cream to the public along with a black hair oil for men whose hair was turning gray. Additionally, in 1910, he invented a curved tooth iron comb for the straightening of women's hair.

Morgan's three businesses prospered so well that he soon built a new home in Cleveland and was driving on its streets in a new automobile. He is believed to be the first man of color to own and operate a vehicle on the streets of Cleveland. While driving one day, he came up with another idea. The development of this idea led to the invention of the modern day traffic light system for highways and streets. This was one of his most practical and well-known inventions. In 1923, the invention of the three-way traffic light was patented. People today drive on inner city streets and more than 42,795 interstate highways across the nation in a safer and more expeditious manner because of Mr. Morgan.

Morgan's traffic light system had only two positions—stop and go . . . red and green, with no neutral position. Amber is the color of the additional light that was attached to the system. Perhaps this was derived from a sociological angle to benefit citizens who may have been color-blind. Today as we have it, there are three positions on the traffic light. With Morgan's improvement to the signal light system, traffic could move in all directions without an officer. When traffic officers were used they often had a hard time keeping the flow of traffic at steady intervals. This inconsistency by the officer caused a delay in the movement of people and vehicles. General Electric Corporation later purchased the rights of Morgan's invention for $40,000.

In 1912, Morgan ingeniously invented the inhalation or breathing machine. The date of this invention certainly shows the extended number of years Mr. Morgan used his talents to make positive contributions to the human race.

In Cleveland, Ohio, on July 24, 1916, there was a fatal explosion in tunnel five of the Cleveland Waterworks. The tunnel was deep in the earth, specifically, under Lake Erie. A large number of men working in the area were trapped with gases, dust and smoke around them. Cleveland city emergency personnel were dispatched to rescue them to no avail. Someone remembered that Morgan had invented a gas mask or breathing machine and Mr. Morgan was contacted at home. He arrived at the scene of the accident with his brother. Immediately the two of them went to work—into the gas and dust of the tunnel, using their gas masks to bring over thirty men to the surface. This rescue mission was a success because of the creation or invention of the gas inhalator.

The gas inhalator allowed Morgan and his brother to breath clean air that was carried in a pocket of the inhalator equipment. They were the only people who were able to enter the gas and smoke filled tunnel to reach the unconscious and deceased workers. This heroic rescue brought his invention from obscurity to the public's attention. At this point, agencies and fire departments within the state of Ohio and around the United States became interested in his great discovery. Orders for the inhalator were received from fire departments and other institutions across the nation.

When it became known that Garrett Morgan was a man of color or Black, many of those who had placed orders for his new invention suddenly began to change their minds about receiving the machines. Morgan had to eventually have a White man to help him make the selling of his traffic lights a success. He further had to pretend that he was a Native American instead of a Black man. The fact that Morgan did not allow the racism in America to hinder his mind and spirit is what made his perfected invention available to various branches of the Armed Services during the first war our country was involved in. The use of his mask during World War I resulted in the saving of many lives.

During the later years of Mr. Morgan's life, he lost his eye sight. Until the end of his life he kept a strong, positive spirit despite the discrimination he constantly encountered. He remained very patriotic to his country despite its rejection of him as a person. He died in 1963 and was unable to attend an Emancipation Centennial Celebration in Chicago, Illinois. He realized that his sight and health were failing, yet he had insisted on being at the celebration.

In the moral fiber of the world, there is a law called the Universal Law of Justice. And flowing out of that fiber and law is the substance which allowed Garrett A. Morgan to keep strong and make so many meaningful contributions to America.

# Chapter 6

# *Elijah McCoy*

## *(1843-1929)*

## *Inventor of the Lubricator of Locomotive Engines*

*"Teach a child his or her roots, and they will take wings and fly away."*

—Unknown

The Smithsonian Institute of Washington, D.C., in recent years, delighted the U.S. populace through exhibiting major creations and inventions of Elijah McCoy in several states from California to Florida. The public response to the exhibitions were all positive.

Mr. McCoy was born in May of 1843 in Ontario, Canada. He was an American citizen in the sense that both of his parents were born in the state of Kentucky. They escaped to Canada, fleeing the institution of slavery. It was as winter was approaching in 1837, via the Underground Railroad System, that McCoy's parents made their courageous departure.

Elijah worked with his father on his farm in Canada until he was fifteen years old. The farm, 160 acres, was given to Elijah's father after his discharge from the Canadian Army. When Elijah reached his fifteenth birthday, his father sent him to Scotland to school. His studies there were in mechanics and engineering. He did well in his studies, staying in Edinburgh, Scotland, for five years. Upon returning to Canada, he remained there for a year or so and then moved to the United States.

When McCoy arrived in the United States in his early twenties, around 1865, the President of the country was Andrew Johnson. President Johnson's administration was highlighted by the Thirteenth Amendment to the Constitution, abolishing slavery, and the Reconstruction Act.

The start of the work years for McCoy within the United States began as legalized slavery was ending. This made it difficult for him to get a job in accordance with his education and background in mechanics and engineering. The remnants of slavery that continued to exist in America kept him from getting a job equal to his abilities and training. Therefore, his first job was on a train as a fireman. He was disappointed and depressed about his predicament, but still accepted the job with the Michigan Central Railroad.

His duties as an employee of the Michigan Central Railroad were to provide the items needed to give the trains energy to move. The train's mobility was generated by the heat of a wood furnace. He kept the stoker of the train furnished and filled with wood. In the process of doing his job, he noticed other workmen on the train standing on the running board to keep its wheels greased. McCoy felt that there should be a better way to keep the wheels of the train lubricated. His inventive mind started to work overtime!

Elijah McCoy started his own business, a machine shop, in Ypsilanli, Michigan. Here he began to consistently experiment on a way to grease or lubricate the wheels of the train without going outside of it onto the running board. After two years of work in this area, he had created a Lubricator serving this purpose. On July 2, 1872, he had his new invention patented. The Lubricator worked on the engines as he had hoped it would. He continued, though, to work at making it more effective and to perfect it.

Despite the fact that he had already invented one creation to serve as a mechanism to lubricate the train wheels, he became a greater winner by not stopping until it was perfected. In this process of refining his invention, he created several other lubricators in varying forms which operated more efficiently. He patented an additional six lubricators.

Since evil thoughts lingered in some men's spirits against this man of color and his accomplishments, some train workmen objected to the use of the lubricator mechanisms because they were made by a Black inventor. He was also known to have been called the "N" word . . . and his invention called a "nigger oilcup."

However, McCoy's invention of the Lubricator Oilcup was eventually considered a progressive and highly valued instrument that helped to advance the railroad industry considerably. In addition, its presence would override the foul thoughts in certain men's minds about Elijah's race. Within a short number of years, many railroads in the United States and around the world were using the Lubricator Oilcup. He even had to instruct train owners and workmen on how to install and use the Lubricators. Between the years of 1872 to 1915, most locomotives in this nation and in foreign countries were in need of

McCoy's services. It is known that many of his inventions were patented in the countries of Germany, Russia, Great Britain and others.

When the change of train engines shifted from steam to Superheaters, McCoy, once again responded to the needs of the new type of engine by creating a Graphite Lubricator. The Superheater Engines used extra large amounts of steam. This additional output of steam required a different type of lubricator. The Graphite Lubricator was patented by McCoy in 1915. The Graphite Lubricator was invented at the McCoy Manufacturing Company, which Elijah McCoy organized in Detroit, Michigan. Here, he also began to sell his new invention.

Before the creation of his Graphite Lubricator, the railroad industry had entered a troubled period over the Superheater Engines. The problem that developed by the use of the Superheater Engines was made clear by Mr. Kelly who wrote an article in the *Engineer's Journal*: *"Our problem from trying to lubricate cylinders of Superheater Engines is not so much due to lack of oil to withstand the heat of the cylinders, as to lack of some way to supply the oil we have with some regularity while the engine is working."*

In a book entitled, *9 African-American Inventors,* by Robert C. Hayden, there is described the system that made the Graphite Lubricator work. McCoy's new lubricator used a solid substance called graphite as the lubricant. Graphite is a form of the element carbon, and is a basic substance found in the lead of an ordinary pencil. The properties of lead, which are softness, smoothness and greasiness made it an excellent lubricant. If a lead substance is mixed with water or oil, it makes a good lubricant. This is the method used by McCoy to finalize his Graphite Lubricator invention.

By the mid 1920's, because of the genius of his mind, McCoy had grown in stature and was known around the entire world. He was called upon by large industries to do consulting work.

In Detroit, Michigan, where he spent a majority of his adult years, he spent time with youth of various neighborhoods. He spent countless hours in counsel with them because he believed they could do the same as he had done if they would only apply their hearts and minds to the task.

Elijah McCoy remained strong physically and mentally until he had reached his upper seventies. He remained a very proud man because of the fifty-seven inventions he had coined, and that had become known as assets to the American society and the world. Mrs. Elijah McCoy died a few years in advance of Mr. McCoy, who passed in 1929. Elijah McCoy's eighty years on earth continue through the creativeness of his thoughts and inventions. He will live eternally as present and future generations build upon the progressive legacy he has left. The brightness of the future is best seen through a listing of many of McCoy's discoveries and inventions cataloged below:

### Other Great Inventions and Discoveries by Elijah McCoy

| Invention | Patent No. | Patent Date |
|---|---|---|
| Lubricator | 139,407 | May 27, 1873 |
| Lubricator | 255,443 | Mar. 28, 1882 |
| Lubricator | 261,166 | July 18, 1882 |
| Lubricator | 320,379 | June 16, 1885 |
| Lubricator | 357,491 | Feb. 08, 1887 |
| Lubricator | 383,745 | May 29, 1888 |
| Lubricator | 418,139 | Dec. 24, 1899 |
| Lubricator | 465,875 | Dec. 29, 1891 |
| Lubricator | 472,066 | Apr. 05, 1892 |
| Lubricator | 610,634 | Sept. 13, 1898 |
| Lubricator | 611,759 | Oct. 04, 1898 |
| Oil Cup | 614,307 | Nov. 15, 1898 |
| Lubricator | 627,623 | June 27, 1899 |

| | | |
|---|---|---|
| Lubricator for Steam Engines | 129,843 | July 02, 1872 |
| Lubricator for Steam Engines | 130,305 | Aug. 06, 1872 |
| Steam Lubricator | 146,697 | Jan. 20, 1874 |
| Ironing Table | 150,876 | May 12, 1874 |
| Steam Cylinder Lubricator | 173,032 | Feb. 01, 1876 |
| Steam Cylinder Lubricator | 179,585 | July 04, 1876 |
| Lawn Sprinkler Design | 631,549 | Sept. 26, 1899 |
| Steam Dome | 320,354 | June 16, 1885 |
| Lubricator Attachment | 361,435 | Apr. 19, 1887 |
| Lubricator for Safety Valves | 363,529 | May 24, 1887 |
| Drip Cup | 460,215 | Sept. 29, 1891 |

# Chapter 7

# *Jan E. Matzeliger (1852-1889)*

## *Inventor of the Shoe Lasting Machine*

*"History does not repeat itself except in the minds of those who do not know history."*

—Unknown

It was nearly 100 years after having invented the shoe lasting machine that Jan Ernest Matzeliger was remembered by the National Association for the Advancement of Colored People (NAACP) on May 16, 1967, for his nineteenth century creation.

It was in the small town of Lynn, Massachusetts, that the NAACP asked for the celebration and called it the Jan E. Matzeliger Day. Jackie Robinson, the first Black to play baseball in the Major Leagues, spoke on that occasion. Jackie, as he was affectionately called, is still remembered with high esteem by American Blacks because of his natural affinity for the game of baseball, and for his willingness to gracefully endure the ugly and shameful attacks upon him by some fans attending games at Ebbets Field in Brooklyn, New York. Jackie was a dark complexioned man with great skill in running, hitting and baseball glove handling. Yet, many of the fans would yell from the stands, "you black nigger" and would throw black cats onto the field.

It was years later when Jan Matzeliger came of age and started to look for work that he began to experience the same kind of insults that continued to be encountered by Robinson as he entered professional baseball in 1947. The reality of the deep-seated racial hate and discrimination within the society became a fact that was hard to face as he moved into adulthood.

Historically, in 1877, Jan Matzeliger had apparently become informed that Lynn, Massachusetts, was the shoe manufacturing capital of the New England region. He decided to move there from Philadelphia, Pennsylvania, where he had been living for a few years. It was in the winter of the year of 1877 that he moved to Philadelphia. He moved there as a poor and undernourished young man.

He spoke very little English, for he had been born in a foreign land. He was born in 1852 in the South American Country of Surinam (Dutch Guiana). His father was a native of that country. His mother was a native of West Africa, and married his father who was of Dutch descent. His father worked as an engineer for the government of his native land. It is suspected that his mother, who was from the shores of West Africa, was part of the slave trade. Some 300,000 slaves, over a period of nearly 180 years, were brought to that country against their wills.

Jan Matzeliger may have had some innate ability for learning, and thus acquired great knowledge in operating machines, since his dad worked in the field of engineering. He was hired by his dad at the age of ten to work in his machine shop. It was observed at this time that he seemingly had a knack for learning the intricacies of machines and how they worked.

He worked with his father for eight or nine years and then decided to travel to see the world. At age nineteen, like young men of today who join the military services to experience and see the world, he joined and then began to work for an Indian merchant shipping company.

After two years at sea, his ship docked in Philadelphia, Pennsylvania, and he decided to leave his job and remain in Philadelphia. He started to work at odd jobs dealing with mechanics, and developed into a very effective man using his thoughts and hands to create things. He used his spare time and evenings to study mechanics and to improve his use of the English language.

It is thought that while Jan lived in Philadelphia, he began to develop strong Christian beliefs. Upon his arrival in Lynn, Massachusetts, he began to visit various churches. Before he had settled into a specific church, he had been rejected from several White congregations because of his race. This type of theological contradiction from some of the church people in Lynn discouraged him somewhat, but it did not lessen his faith in God. He received rejection and hostility despite the fact that he was a religious man. He wore a small medal that read "safe in Jesus" in the lapel of his coat.

In a sense, his movement to a new environment, another state, apparently was guided by a brighter and invisible theological spirit that prevails in the universe. For this slender, erect young man, known for his sense of humor and friendly nature, carried these positive traits with him despite all of the insults and rejections he had already encountered in his life.

During his first year in Lynn, Massachusetts, Matzeliger began to work in a shoe factory. At the shoe factory, he was surprised to learn that making one shoe took such a long period of time. The lesson he learned was that hand lasting was a slow process. It took a considerable amount of time to pleat the leather and fit the uppers to the soles. With his creative and energetic mind, Matzeliger could not conceive that a machine had not already been invented to speed up the manufacturing of shoes. He quietly and quickly decided within himself to accomplish the creation of such a machine.

Matzeliger rented a room over an old west Lynn Mission. He started to design models of his ideas, using old cardboard boxes, scraps of iron and wood, and improvised tools. He remained creative and industrious and he never stopped working and studying his interests and his trade. He bought books on physics and mechanical science and the necessary instruments to do his drawings. His trial and error experimentation continued for several years. He was fully on his way to producing a machine that would manufacture shoes. By using an old forge abandoned by a blacksmith, he was able to shape the needed gears and cams for his machine. After an extended period of working in his rented shop, a rough model began to take form.

Jan worked diligently at making drawings and models of machine parts and showed how they could work together. With experience he had acquired at the factory, he began to make his own hands move the right way to produce the shoes. He became poorer since he used his earnings from his work at the factory to purchase materials for the machine he was building. He denied himself food and nourishment to fund his dream. It is known that often he would eat nothing but cornmeal mush for his daily meal.

The word got out about the work Jan was doing in his shop over the Mission. His first model was very crude. It was made from scrap items he had collected from various places. When some people from the public saw it, they laughed at it. He listened to their criticism but never lost hope. When Jan decided to make the model of his shoe machine out of metal, he acquired a small space in the factory where he worked in the day time. Well into the night, he worked hard at creating the type of metal model he desired. Molding, shaping, reshaping, filing, machining and fitting, Matzeliger put his heart and soul into this project.

At last he had a model of the machine that he thought was ready to be patented, although he had no cash to put the final touches on the invention. He became more convinced that he had an excellent project. Then investors began to approach him about buying his idea. Two wealthy men in Lynn came to Matzeliger and offered to underwrite or finance his invention. For providing this service, Matzeliger agreed to give them two-thirds ownership of the machine. When he described his invention in writing to the patent office in Washington, D.C., they did not believe that such a machine had been created! A representative from the patent office even went to Lynn to examine the "Lasting Machine" model. Finally on March 20, 1883, Mr. Matzeliger was awarded patent number 274,207, for his invention.

Matzeliger's invention was an immediate success. Its adoption by the shoe manufacturing industry created thousands of new jobs, where before only a few master craftsmen could be utilized. With a faster manufacturing time for shoes, the prices were cut and wages were doubled. It is known that thousands of White immigrants left their European poverty to come to work in the prosperous shoe industry spurred by a Black inventor. The exporting of shoes increased from one million pair to eleven million pair within a year. Within a few years, Lynn, Massachusetts, became the world's largest shoe manufacturing center.

Other patents obtained by Matzeliger were for items such as a mechanism for distributing tacks, patent no. 415726, November 26, 1899; nailing machine, patent no. 421,954, February 25, 1892; and even a tack separating mechanism, patent no. 423,937, March 25, 1892.

Mr. Matzeliger did not live long enough to really see the full results of his creativity and hard work. He had worked and driven himself so hard that his resistance broke down. He became increasingly weak and exhausted from the strain of pushing himself for long hours. His health became worse and he was later bedridden. Finally, Matzeliger's illness was diagnosed as tuberculosis. His medical doctor had him carried to the hospital in Lynn, Massachusetts. He spent the last days of his life in a medical facility where he was cared for until his death on August 24, 1889. He was a young man of thirty-seven years of age.

Matzeliger made out a will a few months before becoming deathly sick. In it, he remembered the people who had been close, caring and kind to him. He left shares of stock to the doctor and medical personnel at the hospital of Lynn, Massachusetts. To other close friends, he left his Bibles, paintings done during his sickness, and the tool he had used while he was doing his initial drawings and sketchings of his inventions.

The North Congregational Church of Lynn was the greatest recipient of his possessions. This was the church that accepted him as a person and member of its congregation after several others had openly rejected him because of his race. At the time, $10,000 was a considerable amount of money, and this was the amount of money that he left to the church. The congregation used his gift to pay off its debts. North Congregational Church later merged with another church and renamed itself First Church of Christ. The new congregation remembered Jan in a Sunday morning service on September 8, 1968. In the bulletin for that service, a statement was written regarding his contributions to the New England region. It read as follows:

> *Jan Ernest Matzeliger's invention of the shoe lasting machine was perhaps the most important invention for New England. His invention was the greatest forward step in the shoe industry. Yet because of the color of his skin, he was not mentioned in the major history books of the United States . . . We are not honoring Matzeliger because he gave the church money, but because he is a hero with whom the American people can identify. Mr. Jan Ernest Matzeliger, the young students of school age in America today will also draw upon your rich heritage*

*and gifts to America and strive like you with all our unrelenting strength to make our contributions also to the nation.*

Another inventor named Lewis Latimer, wrote a poem and entitled it, *Unconquered and Unconquerable.* The spirit and drive of many Black inventors during this time truly could have been deemed unconquerable. With initiative, independence, resourcefulness and tremendous talent and will, these brilliant minorities overcame the odds.

# Chapter 8

# *Norbert Rillieux*

## *(1806-1894)*

## *Inventor of the Sugar Refining Machine*

*"The Negro now stands at the crossroads of human dignity . . . If he goes backward he dies: If he goes forward it will be with the hope of a greater life . . . We are either on the way to a higher racial existence or racial extermination . . ."*

—Marcus Mosiah Garvey

Today in America, in many instances, sugar and its uses are taken for granted. It is estimated that each person in the United States consumes 100 pounds of sugar per year! Before it reaches our tables, factories, stores, hospitals, restaurants and schools, it goes through a refining process. And it is the method used to refine the sugar that makes it into the more acceptable form for use. In 1840, Norbert Rillieux was the man and inventor of the device known as the Sugar Refining Machine that caused the processing of sugar to be less complicated and more readily available for consumption.

Mr. Rillieux was born in March of 1806 on a plantation in New Orleans, Louisiana. The details of his life are sketchy and complex. It is known that he lived most of his adult life and attended school in France. However, this is somewhat of a puzzle because we know that he was born on a plantation in the southern part of the United States. Rillieux's father, a White Frenchman named Vincent, was a master and chief engineer on a plantation in Louisiana. It is believed that Norbert's mother was a slave at some point and time on this specific plantation. Norbert's birth record indicates that he was born free. His mother must have gained her freedom before his birth, otherwise Rillieux's record would have shown him as being born a slave.

In his early childhood years, Norbert's father considered him a very bright child. In light of this belief, Vincent Rillieux sent his son to France to be educated. This is perhaps the very reason he spent so much of his life in a foreign land. Indeed, he started to excel early on in his studies in France. His intelligence and innate abilities were in the field of science and engineering. Norbert was energetic and a fast learner who grasped materials so easily that he eventually began to teach and instruct other students before he was twenty-five years old. In 1830, he was teaching applied mechanics in Paris, France. It was during this time in France that he thought of the formation of a machine that could process sugar. Norbert Rillieux's vision intensified as he began to search for a solution to the problem of evaporating the water from the sugar juice in such a manner that the sugar would become crystals.

At an early stage in his manhood and teaching career, he had published articles relating to his understanding and work of steam engines. About the age of twenty-four, he already had several inventions to his credit. Now with accelerated efforts, Norbert's energy turned to developing his idea of a sugar processing machine. Still living in France, he was unable to get French manufacturers interested in investing in his idea of a multiple-effect evaporator. He was unsuccessful getting French machinery manufacturers to even entertain thoughts of helping him to experiment with his ideas and build a device.

There is a saying that "a man's name precedes and goes ahead of him." Thus, Norbert's name as an exceptional engineer reached the New Orleans region of the United States. A businessman who was building a new sugar plant called Rillieux and invited him to work in America. Rillieux had become distressed and disappointed because he could not convince Frenchmen to invest in his idea, so in 1830, he accepted the invitation and went to Louisiana. The work experience for Norbert in New Orleans did not flourish. His father, Vincent, had some differences with the owner of the sugar refining plant, and Rillieux resigned in order to keep peace between his father and the plant owner.

There is an old axiom, "a quitter never wins and a winner never quits." This can be applied to Norbert's life because he had several other negative working experiences in the machinery world before he finally decided to try selling real estate. He did succeed in the real estate world! He made large sums of money, but in 1837 he lost it all when a bank failed.

Other attempts to get his multiple-effect evaporator patented also failed. He had failure after failure, but he refused to give up on his idea until in August 1843, he was finally awarded a patent for his invention. In every application made for a patent, a description must be submitted by the inventor; Mr. Rillieux's description of his evaporator read:

> *"A series of vacuum pans, or partial vacuum pans, have been so combined together as to make use of the vapor of the evaporation of the juice in the first to heat the juice in the second, and the vapor from these to heat the juice in the third which later is connected with a condenser. The degree of pressure in each successive one being less . . . The number of syrup pans may be increased or decreased at pleasure so long as the last of the series is in conjunction with the condenser."*

The main thrust of Rillieux's creation was based on one central thought: That the steam created by regular water was used for heating sugar juice and thus evaporated water from the juice. The vapor that came from the juice upon evaporation was used to evaporate another pan of juice. This simple process of heated water resulted in steam which caused the sugar juice to become crystallized. Understanding this formula and translating it into a mechanical device resulted in the actual invention of the sugar processing machine.

As fate would have it, in 1845, Norbert met a Mr. T. Packwood in New Orleans. Mr. Packwood was a sugar manufacturer who owned and operated a plantation. He invited Rillieux to come to his plantation and use his new invention to refine his sugar. The multiple-effect evaporator or the "Rillieux Machine" worked flawlessly and performed on Packwood's plantation with total success! Those who were close to the sugar refining business commented that Rillieux's evaporator was the first workable machine of its kind in the world. Rillieux's 1845 system was considered by the experts in the sugar industry to be a revolutionary factor in their business. It produced an excellent grade of sugar at reduced costs and many factories in the Louisiana region started using it.

Norbert Rillieux's new method of processing sugar replaced the old rustic way of evaporating sugar juice in open kettles. His new techniques introduced in the sugar business caught on also in Cuba and Mexico. His stature grew until he became the most popular engineer in that region of the United States and many parts of the world.

The next ten years, from 1845-1855, was a time of glory and fame for Mr. Rillieux. In the past, on many of the plantations, slaves used long handled spoon-like gadgets to move the boiling sugar from one steaming hot kettle to another. Now with Rillieux's new creation, valves were used to transfer the hot juice from one place to another. This method reduced the amount of labor and the amount of steam used.

It is known that Rillieux left Louisiana to return to France in approximately 1855. He had built quite a reputation for his intellectual and mechanical expertise. Yet, it is thought that he returned to France because of the institutionalized experiences of racism against him. He understood clearly that he was treated differently because of the color of his skin or racial ancestry.

In the United States, just before the signing of the Emancipation Proclamation by President Abraham Lincoln, people of color in Louisiana could not walk the street without receiving special permission. This rule even applied to slaves that considered themselves free. Every "free slave" was subject to the worst kind of harassment and ridicule. A Black person traveling in the region could not travel alone unless accompanied by a White person. This time in U.S. history, 1855, is considered the Pre-Civil War years and conditions were horrible and taxing for persons of African descent in America. It is believed that this is why Rillieux finally decided to return to France. Maybe as he was struggling with his decision to return to France or to remain in America, a stanza of a great song flashed through Rillieux's being . . . "America, America, God shed His grace on thee. And crowned thy good with brotherhood, from sea to shining sea."

Nonetheless, he did return to France. The degree of

racial disharmony in France was less, but upon his return there he encountered another serious problem. Some of his closest friends believed that Rillieux did not want to leave the United States to return to France. He did so because of the prejudice and racial hatred he, and many other Blacks, faced daily.

It was during the late 1850's when Rillieux returned to France and encountered some real trouble concerning his inventions and patents. He discovered upon his arrival that a German man who had worked in Philadelphia for a firm that constructed Norbert's original multiple-effect evaporator had copied the designs of it. This created a major difficulty for Rillieux. This event made it truly hard to determine who was due the credit for the invention in question. The evaporators, however, that had been built in France from the stolen copies did not operate as efficiently as those built in the United States. Apparently they did not function correctly because they had not been constructed properly. The misuse of the stolen copies of the evaporator created a bad name in France for their well respected inventor, Rillieux. As a result Rillieux was unable to find a sugar manufacturing company interested in examining his original machine.

The whole episode had so damaged his reputation and had discouraged him to the extent that he lost interest in the very machine that had modernized the sugar processing industry. For ten or more years he shifted to another profession. He entered the field of archeology. In 1880, a businessman from New Orleans traveled to France and was surprised to find Mr. Rillieux working for a library translating the Egyptian language.

For some reason, at age seventy-five, Mr. Rillieux again focused on his first love—engineering and machinery. He thought of how his evaporator might extract sugar from the sugar beet plant and with renewed energy sought to get a patent for the manufacturing of sugar from the plant with his evaporating system. He did receive a patent in 1881. Many of the sugar beet businesses accepted his offer and Rillieux cut their fuel cost in half as a result of his machine being used.

In France, many of the seasoned experts refused to give him all of the credit earned for his invention of the multiple-effect evaporator. He finally retired from this work that he was best suited for—still somewhat dismayed. A close associate of his remarked that as he approached his senior years he was still an active and alert man, but said, "He died more of a broken heart than from the weight of his years."

Dr. Charles Brown, sugar chemist of the United States Department of Agriculture said, "Rillieux's invention is the greatest in the history of American Chemical engineering." Some twenty years after Rillieux's death, in 1914, a Dutch sugar expert contacted many in the sugar industry from around the world to honor and remember his work. The list included the President of the International Society of Sugar Cane Technologists. There were a total of thirty-eight contributors to this memorial celebration. And in the state of Louisiana, there is a plaque in honor of Rillieux in the museum. The inscription reads:

*"To honor and commemorate Norbert Rillieux born on March 18, 1806, New Orleans, Louisiana, and died at Paris France, October 8, 1894. Inventor of the Multiple Evaporator and its application into the sugar industry. This tablet was dedicated in 1934 by corporations representing the sugar industry all over the world."*

For hundreds of years, before Rillieux's invention, the old fashioned way had been applied in the manufacturing of sugar. His discovery was not just switching to a mechanical device, but it completely changed the sugar industry.

# Chapter 9

# Frederick McKinley Jones (1892-1961)

## Inventor of the Removable Cooling Device

*"Each generation must out of relative obscurity discover its mission, fulfill or betray it . . ."*

—Frantz Fanon

Frederick McKinley Jones was born during the time when William McKinley was elected the twenty-fifth President of the United States of America, which was from 1897 to 1901. President McKinley was a Republican and his administration was highlighted by the Spanish American War and the Open Door Policy. It is thought that Jones' middle name can be associated in some way with the name of President McKinley.

Jones was an orphan at age ten. His mother died when he was an infant. His father died when he was just nine years old. His birth place was in Cincinnati, Ohio.

After the death of his father, he left his home and traveled to Kentucky where a priest named Father Ryan cared for him. He lived in the rectory and odd jobs became his responsibility. He left Kentucky and Father Ryan when he was sixteen years old, as he decided to seek his own independent way. This required that he drop out of grammar school in the sixth grade in order to take care of himself. Between the ages of ten and sixteen he worked various odd jobs, i.e., being water boy on construction sites and as a pin setter at bowling alleys. In early experiences, he became intrigued with gasoline engines and the details of other types of machinery. It was observed in his early working years that he had an innate ability and talent for the understanding of motors. At age sixteen, he had grasped enough information about the workings of motors to be a journeyman mechanic. In just a three year period he had become a shop foremen, specializing in building race cars from the frame up.

However, his upward mobility that led him to become a top notch engineer and inventor was a continuous yet difficult climb. In addition to his intellectual capabilities he apparently had mental, emotional, and spiritual strengths to succeed against the failures and difficulties of life. Even at the death of his father and mother, this natural phenomenon did not deter or redirect his energies to become a successful man and inventor of many practical, useful, and meaningful items that have lifted the standard of living for people of all ethnic groups in America and around the world. His inventions are listed below in order that the public, students and non-students alike, will grasp a clear picture of the magnificent character and ability of this man. This listing is also being done to show the power that can be in the human spirit and mind to succeed with an unwillingness to give in to life's circumstances:

| Invention | Patent No. | Patent Date |
|---|---|---|
| Ticket dispensing machine | 2,163,754 | June 27, 1939 |
| Air Conditioning Unit (the refrigerated truck) | 2,475,841 | July 12, 1949 |
| Method for air conditioning (refrigerated boxcar) | 2,696,086 | Dec. 07, 1954 |

| | | |
|---|---|---|
| Method for preserving perishables (refrigerated boxcar) | 2,780,923 | Feb. 12, 1957 |
| Two-cycle gasoline engine | 2,532,273 | Nov. 28, 1950 |
| Two-cycle gas engine | 2,376,968 | May 29, 1945 |
| Starter generator | 2,475,842 | July 12, 1949 |
| Two-cycle gas engine | 2,417,253 | Mar. 11, 1947 |
| Means for thermostatically operating gas engines | 2,477,377 | July 26, 1949 |
| Rotary compressor | 2,504,841 | Apr. 18, 1950 |
| System for controlling the operation of refrigeration units | 2,509,099 | May 23, 1950 |
| Apparatus for heating or cooling the atmosphere in an enclosure | 2,526,874 | Oct. 24, 1950 |
| Prefabricated refrigerator construction | 2,535,682 | Dec. 26, 1950 |
| Refrigeration control device | 2,581,956 | Jan. 08, 1952 |
| Methods and means of defrosting a cold diffuser | 2,666,298 | Jan. 19, 1954 |
| Control device for internal combustion engine | 2,850,001 | Sept. 12, 1958 |

The patent office in Washington, D.C., has a listing of all Mr. Jones' inventions as required by law. Along with each invention, a description of the mechanism is also required. A detailed account of his most famous invention, the removable cooling device which was patented on July 12, 1949, read:

*My invention relates to a removable cooling unit for compartments of trucks, railroad cars and the like, employed in transporting perishables and to a method of cooling such compartments and has for its object to provide a simple and compact, self-contained cooling unit positioned at the top of said compartment and combined with air flow passages which produce a vortex of cold air flowing about the walls of the compartment and returning from the center of the compartment. Perishables such as fruits, meats, vegetables, and the like are transported in what are known as refrigerator cars by rail and, to a greatly increased degree at the present time in trucks. This transportation, taking place as it does over long routes which in the summertime are at high temperatures throughout and even in wintertime may be in part at high temperatures, required artificial cooling in order to preserve said perishables in suitable condition for use as food. In the case of this, it is not true of trucks where the necessary limitations of their use call for cooling means relative of refrigerator railroad cars; heavy cooling means such as large ice compartments or large and heavy refrigerating plants can be practically employed. This is not true of trucks where the necessary limitations of their use call for cooling means relatively low in weight and so positioned as to take up as little as possible of the space within the transporting compartment.*

*It is a principal object of my invention, therefore, to provide a cooling unit small in size and weight, and positioned together with the air-conducting passages, so as to occupy substantially none of the storage space within the vehicle compartment. It is further an object of my invention to provide a unit which shall be mounted in the front wall of the compartment partly outside and partly inside and having its top adjacent to the top wall of the compartment.*

Frederick M. Jones, a grammar school dropout with consistent drive and motivation, revolutionized the transportation industry with this invention. The marketing patterns of the total nation changed as a less expensive transportation of frozen foods developed around his idea of making refrigerated vehicles a reality.

There were at least two branches of the United States Armed Services that also benefited from his invention.

His refrigeration systems were utilized for all Army and Marine Field Kitchens. During World War I, he gave the Army an important element in the area of food economics—the refrigerated boxcar.

During World War II, Jones gave the military services a great gift in the portable cooling units. In the South Pacific during this particular war, Army hospitals and battlefields used refrigerated units designed by Jones for this specific purpose. Those units were used for keeping blood serum for transfusions and other medicines at a desired temperature. Jones' portable units were used by Army airplanes and helicopters and were flown into deep jungle territories of the South Pacific islands. Also, other units were utilized for flying wounded soldiers from the front lines and others back home by way of the Pacific Ocean. If the need arose, the units could produce heat. The first portable x-ray machine designed by Jones also proved to be very useful in treating the wounded of the war stricken areas.

In his sixty-nine years, Jones was awarded a total of sixty patents. Of this number, forty of them were for his refrigeration systems. Jones held patents for such things as the air cooling units, self-starting gasoline engines used to turn the units on and off, the reverse cycling mechanism for creating heat or cold devices for controlling air temperature and moisture and parts for all of the above equipment. In America, Frederick McKinley Jones became an authority in the field of refrigeration. In the mid 1940's he was elected to membership in the American Society of Refrigeration Engineers. Though he was a grammar school dropout, college professors, engineers and scientists were anxious to work with him and learn from his wisdom. He was also a consultant to the United States Defense Department and to the Bureau of Standards.

In 1961, Frederick McKinley Jones passed from the human scene. At the time of his death, he lived in Minneapolis. The inventions he created during his lifetime served not only people in his home state and town, but reached and served people around the entire world. Of the many contributing people within our global society who work out front in jobs, positions and professions that are very visible, Frederick functioned behind the scenes, even unknown to many of the people in the United States. However, what he quietly produced to serve humanity will always have far reaching effects upon our social order.

Chapter 10

# Andrew Jackson Beard (1849-1921)

## Inventor of the Automatic Railroad Car Coupler

*"Intolerance can grow only in the soil of ignorance; from its branches grow all manner of obstacles to human progress . . ."*

—Walter White

Andrew Beard was born in the extreme southern region of the United States. He was born a slave on a plantation in 1849 in Jefferson County, Alabama. His emancipation, or freedom, came at the age of fifteen and he married the next year. In 1849, the year of Beard's birth, Zachary Taylor was elected the twelfth President of the United States. President Taylor served from 1849-1850. He was a member of the Whig Party, a political party that preceded the Democratic and Republican parties. President Millard Fillmore, the thirteenth president followed President Taylor in that office. President Fillmore was also a Whig member and served as President from 1850-1853. His administration is remembered fro the Fugitive Slave Act and for California becoming a state.

The Fugitive Slave Act dealt with the question of slaves fleeing, especially from the deeper southern states to northern parts of the country in the quest for their freedom. The fugitive slave law was an act of Congress, passed in 1793 and 1850, providing for the surrender and deportation of slaves who escaped from their masters and fled into the territory of another state, generally a free state. Andrew Beard's emancipation at an early age in his life is what obviously gave him the latitude to develop his talents to become a carpenter, blacksmith, railroad worker, businessman, farmer, and inventor.

Beard was a farmer in the Birmingham, Alabama, area for about five years. During his farming experiences, he once visited Montgomery, in 1872, with fifty bushels of apples drawn by oxen. He recalls, "It took me three weeks to make the trip. I quit farming after that." Instead, he built and operated a flour mill in Hardwicks, Alabama. He began to think about developing a plow during this era of his life. When his idea had come to fruition, in 1881, he had it patented. In 1884, he sold his plow for the sum of $4,000.00. On December 15, 1887, Beard invented another plow and sold it for $5,200.00. When the sales of these inventions consummated, he used the profits to start a real estate business. His efforts in this area were successful and he quickly accumulated $30,000.00

This gifted African-American also had in his mental capabilities the feel and understanding of engines and other types of machinery. While he worked for various railroad yards carrying out assorted tasks, it occurred to him that the many injuries to others who labored there did not have to be. Beard was often heartbroken over the tragic accidents that happened at the railroad yards. He recalled it usually happened this way: a man would run along the top of a freight train, climb down between cars, hoping to find the hole for a coupling pin to secure and join the cars as they approached each other. The engine of the train would back up, the cars would come together with a tremendous impact, coupling them together. If caught between the cars, workers would be injured with broken bones on various parts of their bodies.

The frequent occurrence of these accidents injuring the railroad laborers served as a stimulant for Mr. Beard to come up with a device or a machine that would work in an automatic manner to link the train cars. Often he was spurred on by the personal impact of a death or injury to a friend, associate or neighbor. He would consequently stay up late at night grappling with making his thoughts a reality—to link the train cars without human contact.

An example of his sincerity was how he would often work at his kitchen table late into the night struggling over how to attach the tail pin to the lock mechanism on the train cars. He never gave up on how to position the operating-rod to "withdraw the lock and release the head." After many hours, days and even months, in the middle of the last ten years of the nineteenth century, he completed a working model of his concept.

In September 1897, he filed a patent application on a car coupling device. The following is an excerpt of the description of his invention submitted to the patent office:

> *To who it may concern:*
>
> *Be it known that I, ANDREW JACKSON BEARD, a citizen of the United States, residing in Eastlake, in the county of Jefferson and State of Alabama, have in-*

*vented certain new and useful Improvements in Car-Couplings; and I do hereby declare the following to be a full, clear, and exact description of the invention, such as will enable others skilled in the art to which it appertains to make and use the same.*

*My invention relates to improvements in that class of car-couplings in which horizontal jaws engage each other to connect the cars; and the objects of my improvents are, first to provide a car-coupler of a simple and cheap form of construction, the coupler assembled in parts adapted to replace any of the pieces as desired; second, to provide a car-coupler having the head and shank constructed in separate parts and pivotally connected by a pin, by which a new head or shank can readily be attached to replace a broken part; third, to provide an automatic car-coupling having a head and side jaw pivotally attached to the shank, the head and jaw adapted to open and close in opposite directions to couple or uncouple the cars . . . The body of the shank is formed to fit the usual car dimensions and is adapted to be attached to the car in the usual manner. Two lugs 2 2′ are provided on the front end of the shank. The lugs extend forwardly from the shank and have pin-holes 3 3′ provided therein.*

*The head 1 is made of cast-steel or other suitable*

*metallic material. The head is provided with slots 5 5′, formed therein to receive the jaw. Slots 6 6′ are also provided in the head to receive the lugs formed on the front end of the shanks. A pin-hole 7 is formed through the head to receive the pin 8, the pin pivotally connecting the head and shank together. The usual pin-hole 9 is formed in the front of the head to connect with the link, the usual slot 10 being provided to admit the use of a link, if desired.*

*The jaw 11 is made of cast-steel or other suitable metallic material, formed as shown, the tail-wings formed thereon being adapted to fit the slots formed in the head. The jaws provided with a pin-hole 12 to pivotally connect the jaw, in connection with the head, to the drawhead shank by the pin 8, as shown.*

*The head-lock 13 is made of suitable metallic material, formed as shown. The lock slides and groves 14 14′, provided in the shank-lugs. A tail-pin 15 is formed on the lock. The tail-pin extends through a bearing 16, provided in the shank. A coiled spring 17 is provided on the tail-pin. The spring pressing against the head of the lock keeps the same pressed forward to engage the concave recess 18, provided in the head and side jaw.*

*The operating-rod 19 extends across the end of the car and is attached thereto by any of the usual methods.*

*Any desired form of cranks can be formed on the ends of the rod. The rod ends or cranks are not shown in the drawings or any particular form claimed.*

*Two rigidly-connected levers 20 20′ extend downwardly from the operating-rod 19. The levers are pivotally connected at their lower ends to the transverse bar 21, connected to the locking device. The rod 22 connects the lock with the head. The rod 23 connects the lock-bar with the side jaw . . . The head and jaw when open allow the draw-heads to come apart and uncouple the cars. The cars, if pushed together, recouple the draw-heads automatically.*

*Having thus described my invention, what I claim as new and desire to secure by Letters Patent, is—1. In a draw-head, the combination with the shank having projecting lugs formed on the front end thereof, of the head having slots formed therein to receive the shank-lugs, the jaw engaging in slots formed in the head, the pin pivotally connecting all the parts together, and the sliding locking device to engage the recesses formed in the head and jaw, substantially as and for the purpose described. 2. In a car-coupling, the combination with the operating-rod having two downwardly-pivotally connected to a transverse bar attached to the locking device, of a rod connecting the head to the lock bar, and a rod connecting the side jaw to the lock bar, substantially as described.*

*In testimony whereof I affix my signature in the presence of two witnesses . . .*

Prior to this great creation, namely the train car-coupling device, Beard even lost a leg as a result of a car-coupling accident. The existence of his new invention eliminated the dangers of hooking railroad cars together by hand and probably saved thousands of lives and limbs of railroad workers. In 1899, he improved his coupling device and later received $50,000 for its patent rights. His improvement to the coupling device in 1899 became the forerunner of today's automatic coupler.

After the issuance of a patent which improved the coupling device in 1899, details and events of the life of Beard became sketchy and mysterious. It is virtually unexplainable where he lived after this time. There is no record available on where he died or his burial place.

Andrew Beard refused to let his lack of formal training in either engineering or metal working deter him from his goal to create something to make conditions better for ordinary working laborers. This steadfast work resulted in providing a great service to the whole of society. We indeed owe a debt to this inventive genius.

# Chapter 11

# Clay S. Gloster, Jr.
## (1962-Present)

## Weighted Random Pattern Generation

*"Up you mighty race, you can accomplish what you will . . ."*
—Marcus Garvey

In this series on inventions, an earlier profile speaks of an inventor, Joe Dudley, who is a native sone of North Carolina.

The presence now in the Tar Heel State is that of Clay S. Gloster, Jr., an electrical engineer and co-inventor of a weighted random pattern generation. His intellectual gifts in the sciences as an African-American inventor and his strength of academia contribute greatly to the entire state. He is a resident of the city of Raleigh, the capital. He is also an assistant professor at North Carolina State University in its department of electrical and computer engineering.

The North Carolina State University campus with its School of Electrical and Computer Engineering is in close proximity to the Research Triangle Park, Duke University in Durham and the University of North Carolina in Chapel Hill, North Carolina. Together, these facilities make this region of North Carolina a giant number one in the United States for technological, medical, and pharmaceutical research and development.

Clay Gloster's work at North Carolina State University in engineering and computer science keeps him involved in interaction with students in research and study from around the world. Research in electrical engineering and computer science led Dr. Gloster to sharing the weighted random pattern generation invention with Franc Bugler, Ph.D., of North Carolina State University, and Gershon Kesom, Ph.D., of Duke University, Durham, North Carolina, who received from the United States Patent Office on August 27, 1991, patent number 05043988 for their co-invention.

The summary abstract and diagram submitted to the

U.S. Patent Office was approved as listed below:

*A high precision weighted random pattern generation system generates any desired probability of individual bits within a weighted random bit pattern. The system includes a circular memory having a series of weighted factors stored therein, with each weighting factor representing the desired probability of a bit in the weighted random pattern being binary ONE. The random bits from a random number generator and a weighting factor are combined to form a single weighted random bit The random bits and weighting factor are combined in a series of interconnected multiplexor gates. Each multiplexor gate has two data inputs, one being a bit from the weighting factor, the other being the output of the preceding multiplexor gate. The random number bit controls the output of the multiplexor. For example, when the control input (random bit) is high, the multiplexor output is the weighting factor bit. When the control input (random bit) is low, the multiplexor output of the preceding multiplexor. The output of the final multiplexor gate in the series is the weighted bit.*

*A weighted pattern generation system is compressed of:*

- *a memory having a plurality multibit weighting factors stored therein;*

• *a random pattern generator for generating a plurality of multibit random bit patterns; and*

• *means, connected to memory and random pattern generation, for combining multibit weighting factors with multibit random bit patterns to form a multibit weighted random pattern.*

The objective in listing the technical language and jargon in the patent summary is for possible assistance and utilization by students, technicians and others in commercial markets.

Increasing the precision of "weighted random pattern generation" can only mean one thing to the everyday consumer, be that consumer government, industry, or individual. It means our initial chance of purchasing a quality electronic product is greatly enhanced. "Weighted random pattern generation" even more so because digital circuits (which make up nearly all consumer products from garage door openers to wrist watches and timers on microwave ovens) will be given a greater opportunity to expose failures or faults with an increase in the input combinations to determine the output of these circuits. This simply means that when any of the above items are purchased, they have already been tested for accuracy of their circuits.

*Clay S. Gloster, Jr.*

Though it is highly unlikely that the everyday consumer will recognize the benefits of "weighted random pattern generation," manufacturers will see an increase in the integrity of the electronic products rolling out of their production facilities, thus insuring the chance of people buying electronic products that work properly the first time they make the purchase.

America will increase its world community standing as an electronic product producer because of the accuracy of the Weighted Random Pattern Generation. This type of precise testing at the factory level was unavailable before this invention.

# Conclusion

*Great Discoveries and Inventions by African-Americans* was initiated, continued, and completed with a dual objective in mind: 1) to list some African-American contributions to the American culture; (2) to show how African-Americans succeeded in science, research and technology in spite of seemingly insurmountable odds which constantly befell their path. They reached the necessary level of self-esteem to achieve and to invent for the benefit of themselves and that of others. Within their character was the ability to reach internally to persevere and to rise above the external circumstances that were constructed to cause them to be failures.

An even broader inspection of their lives will show that a high percentage of the well-known and not-so-famous inventors moved to establish their own businesses to advance the end product and practical usage of their creations. Before the Thirteenth and Fourteenth Amendments were added to the United States Constitution, African-Americans were unable to patent their inventions. Due to Jim Crow laws and other forms of legal discrimination which prevented African-Americans from buying office space, property, obtaining bank loans and getting established, they used much ingenuity in order to stand strong and to circumvent the obstacles that were placed in their path. In many instances they created additions to their residences. These residential additions acted as offices, laboratories or facilities to help create the logistical stability needed to bring the genius of their work on projects to fruition and reality.

The mind-set of these innovators/inventors in becoming entrepreneurs is the height of independence that is needed, urged and encouraged contemporarily for us as a racial group in America, and especially for our youth. As the Twenty-first Century, the new millennium, quickly approaches, our youth must draw upon the energy and insights of their ancestral models in order to expand their horizons and become self-sufficient. These models had the desire, motivation and the initiative to pursue and fulfill their goals. These goal-oriented qualities of our ancestors will become the incentive for a public-policy-forum initiative to nurture and assist our youth in building their own businesses across America.

The following examples typify the genius and success of African-American entrepreneuralism and scientific innovation:

Joe L. Dudley is a Kernersville, North Carolina, resident. In 1997, his ethnic hair care business grossed an income of more than thirty-two million dollars in North America, South America and Africa. Mr. Dudley is a current model and an example to be followed. Our youth must position themselves to be in the presence of such successful entrepreneurs.

Two other examples of great inventions of the past are the telephone apparatus, October 11, 1887, patent #371241; and relay instrument, June 7, 1887, patent #364619. These instruments were invented by an African-American named Granville T. Woods. These devices are currently utilized in the advanced technological computer/telephone systems of today.

The vision and unrelenting minds and spirits of our ancestors to reach within themselves to initiate the desire, and to be motivated above the external conditions that confronted them, is also within our grasp and capability.

At the end of this work, I am more than ever before convinced that as we move into the future, the Twenty-first Century, the information/technological age, we have the ability to draw upon our innate strength to grow, use, and continue to confront the remaining conditions that are present.

*"To whom much is given, much is expected."*

# Appendix I

## Communicative Inventions

**Inventor:** Granville T. Woods
**Invention:** Telephone System and Apparatus
**Location:** Cincinnati, OH **Date:** October 11, 1887 **Patent #:** 371,241
Woods' invention reduced noise and external interference from voice, sound and signals transmitted on telephones. This invention has certainly contributed to today's computerized telephone systems.

**Inventor:** Granville T. Woods
**Invention:** Relay-Instrument
**Location:** Cincinnati, OH **Date:** June 7, 1887 **Patent #:** 364,619
An electromagnetic device, the relay instrument was invented by Woods. This device utilized small current or voltage change to activate switches and other devices in an electrical circuit. His invention improved the construction and sensitivity of inductive telegraphy.

**Inventor:** Granville T. Woods
**Invention:** Railway Telegraphy
**Location:** Cincinnati, OH **Date:** August 28, 1888 **Patent #:** 388,803
Woods invented a railway telegraph that could send or receive a stronger coded electrical signal between railway operators and their trains.

**Inventor:** Granville T. Woods
**Invention:** Apparatus for Transmission of Messages by Electricity
**Location:** Cincinnati, OH **Date:** April 7, 1885 **Patent #:** 315,368
Woods invented an apparatus that made it possible to transmit signal and voice messages over the same line with the same instrument. With minor modifications his invention also enabled voice transmission by Morse Code devices.

**Inventor:** Joseph Hunter Dickinson
**Invention:** Arm for Recording Machine
**Location:** Cranford, NJ **Date:** January 8, 1918 **Patent #:** 1,252,411
Earlier phonographs, called stereo sets today, neither produced very high quality sound or high volume music. Dickinson, the same man whose invention improved the player piano, invented an arm that gave a richer tone to earlier record players. The arm also helped to control the volume for longer distance coverage.

**Inventor:** Lee S. Burridge and Newman R. Marshman
**Invention:** Typewriter Machine
**Location:** New York, NY **Date:** April 7, 1885 **Patent #:** 315, 366
The typewriter is among the most important office machines ever invented. It produces text by reproducing letters and symbols, inscribed onto certain elements, simply by striking desired keys on a keyboard. This concept of keyboard instruction/communication is the most popular means of controlling today's hi-tech equipment and machinery. Computers and word processors use typewriter keyboards; whereas, calculators, audio/video equipment, automated industrial machine, etc., use key pads.

**Inventor:** W.A. Lavalette
**Invention:** Improvements of the Printing Press
**Location:** Washington, DC **Date:** September 17, 1878 **Patent #:** 208,184
Lavalette made overall improvements to the earlier model of the printing press. The printing process of the older models was slow and cumbersome. His improvements made printing faster and easier. It also made the actual printing clearer and easier to read.

**Inventor:** Joseph H. Dickinson
**Invention:** Improvement of the Player Piano
**Location:** Cranford, NJ **Date:** June 11, 1912 **Patent #:** 1,028,996
The player piano was improved by Dickinson when he invented the roll of perforated sheet music. This invention made it possible to automatically move the player mechanism back and forth to play a melody.

**Inventor:** Phillip B. Downing
**Invention:** Letter Box
**Location:** Boston, MA **Date:** October 27, 1891 **Patent #:** 462,093
Phillip B. Downing invented what is common to us today as the letter box. Its unique design not only protects the mail, but has additionally added to the convenience of the public as well as the easy accessibility to the mail carrier.

**Inventor:** William B. Purvis
**Invention:** Fountain Pen
**Location:** Philadelphia, PA **Date:** January 7, 1890 **Patent #:** 419,065
William B. Purvis was the man who invented an instrument that is used in schools, businesses, governmental agencies, etc., commonly known as the fountain pen. This eliminated the cumbersome traditional practice of carrying your ink with you. This new invention has a built-in ink reservoir and automatically feeds ink into the point of the pen. The ink pen also uses refills and is a convenient device in terms of time saving.

**Inventor:** Robert F. Flemmings, Jr.
**Invention:** Improvement of the Guitar
**Location:** Melrose, MA **Date:** March 30, 1886 **Patent #:** 338,727
Robert F. Flemmings, Jr., an African-American, invented the guitar. Music created by this instrument has rocked very famous stars that you know today. Due to Flemmings' improvements in terms of the melodic tones as well as increased volume, we enjoy the relaxing expression of a universal language . . . music. He also added the additional feature of sensitive touch which many rock and roll groups find invaluable.

## Transportation Creations

**Inventor:** Issac R. Johnson
**Invention:** Bicycle Frame
**Location:** New York, NY **Date:** October 10, 1899 **Patent #:** 634,823
Issac R. Johnson invented a bicycle frame that could be broken down into parts so it could be stored conveniently. This invention made it possible for

college students to store their bikes in their dorm rooms. This invention, with some minor alterations is still in use today.

**Inventor:** John W. Butts
**Invention:** Luggage Carrier
**Location:** Springfield, MA **Date:** October 10, 1899 **Patent #:** na
Since the invention of the bicycle, many people from all walks of life have enjoyed riding them. John W. Butts noticed that there was not a place on the bicycle to carry luggage or some other type of gear, so he invented a luggage carrier that enabled bikers to carry a small amount of luggage or some personal items with them.

**Inventor:** Andrew J. Beard
**Invention:** Car Coupling Device
**Location:** Eastlake, AL **Date:** October 10, 1899 **Patent #:** 634,611
Andrew J. Beard's invention of the car coupling device, helped in a big way in revolutionizing the railroad system. His invention made it easier and safer for train workers to hook and unhook railroad cars as needed.

**Inventor:** Granville T. Woods
**Invention:** Tunnel Construction for Electric Railways
**Location:** Cincinnati, OH **Date:** July 17, 1888 **Patent #:** 386,282
Today the underground tunnel is no longer used for electric trains but is used for subways, sewer systems, crude oil transfer systems and passageways for automobiles under large bodies of water. As we progress in the use of super conductors, underground tunnels will become in high demand in the near future.

**Inventor:** Frederick M. Jones
**Invention:** Air Conditioning Unit for Vehicles
**Location:** Minneapolis, MN **Date:** July 12, 1949 **Patent #:** 2,475,841
Frederick M. Jones realized that the air conditioning unit could be used for more than just cooling the house, but it also could be positioned in a truck to help keep produce fresh until it reaches the market to be sold. His invention has helped produce companies to cut losses due to spoilage of produce on its

way to the market place. Jones's invention is still being used today in commercial vehicles.

**Inventor:** A. Beard
**Invention:** Rotary Engine
**Location:** Unknown **Date:** July 5, 1892 **Patent #:** 478,271
Beard's invention, the rotary engine, was instrumental in helping the airplane to fly. They resemble huge fans similar to the fans used in some households. By spinning around at a great velocity, the rotary engine helps the plane to take off and fly.

**Inventor:** William H. Richardson
**Invention:** Improvement of the Child's Carriage
**Location:** Baltimore, MD **Date:** June, 1889 **Patent #:** 405,599
In June of 1889, William H. Richardson made an improvement in the design of the child carriage by putting a maneuverable wheel base on it which resolved the problem of having to pickup and turn the child carriage around in order to go in a different direction. Today there are many different types of child carriages and Mr. Richardson's process of maneuvering the child carriage is still used in a vast majority of these models.

**Inventor:** R. B. Spikes
**Invention:** Automatic Gear Shift
**Location:** Unknown **Date:** December 6, 1932 **Patent #:** 1,899,814
Prior to 1932, all automobiles manufactured used the manual system of gear shifting. The function of the gear system within an automobile is to give the operator the ability to control and guide the automobile in the desired direction at the same time. R. B. Spikes' invention, the automatic gear shift, relieved the operator of the hassle of shifting gears which also made the process of driving much easier. Mr. Spikes' invention is still used today by major automobile manufacturers around the world.

## Safety Devices

**Inventor:** Garrett A. Morgan
**Invention:** Traffic Signal
**Location:** Cleveland, OH **Date:** November 20, 1923 **Patent #:** 1,475,024
At street intersections where pedestrians and vehicles must be moved safely, the police officer was originally the director of traffic. Morgan invented the traffic light to help in this process. His invention has almost entirely replaced the police officer as the traffic director. His traffic light, with its associated red, green, and yellow lights, is the universal standard used in every country on earth today.

**Inventor:** Joseph R. Winters
**Invention:** Fire Escape Ladders
**Location:** Chambersburg, PA **Date:** May 7, 1878 **Patent #:** 203,517
The fire service and firemen are very important to every community. Also important is the speed at which they must do their work. A few seconds can save lives. Winters' invention was an improvement to the original fire escape ladder. To make rescue faster, his invention made it possible to raise the ladder to higher floors without removing it from the fire truck.

**Inventor:** J. B. Rhodes
**Invention:** Water Closet
**Location:** Unknown **Date:** December 19, 1899 **Patent #:** 639,290
Rhodes invented the commode/toilet used in the privacy of our homes today. Before this invention, people had to go outside of their homes to use the outhouse.

**Inventor:** Garrett A. Morgan
**Invention:** Breathing Device
**Location:** Cleveland, OH **Date:** October 13, 1914 **Patent #:** 1,113,675
Morgan's invention, the breathing mask, can be used by anyone who works where there are harmful fumes or dust. His invention was particularly important to firemen because they could re-enter a burning building to save lives without fear of suffocation by smoke or harmful fumes.

**Inventor:** William F. Burr
**Invention:** Switching Device for Railways
**Location:** Agawam, MA **Date:** October 31, 1899 **Patent #**: 636,197
Train tracks are not like roads where the driver can change direction by turning the steering. Trains make turns when they are shifted from one track to another. Burr's special shifting device helped to easily shift from one track to another. This device allowed the train engineer to move his train easily from the main track to a branch track.

**Inventor:** Greenville T. Woods
**Invention:** Automatic Cut-Off Switch
**Location:** Cincinnati, OH **Date: January** 1, 1889 **Patent #**: 395,533
This was an invention which we know today as the light-switch. It was, and still is a convenient way to turn lights on or off from practically anywhere in the room.

**Inventor:** Humphrey H. Reynolds
**Invention:** Safety-Gate for Bridges
**Location:** Detroit, MI **Date:** October 7, 1890 **Patent #**: 437,937
Bridges are sometimes built over waterways used by ships. Quite often the bridges are built too low; therefore, a special type of bridge, called the draw bridge opens to let ships pass. Reynolds invented the safety gate to help ensure that land traffic would not attempt to use the bridge while it is raised.

## Life Chores Made Easier

**Inventor:** John Hunter Smith
**Invention:** Lawn Sprinkler
**Location:** Unknown **Date:** May 4, 1897 **Patent #**: 581,785
John H. Smith invented a lawn sprinkler that made the task of watering lawns, flower beds, and gardens easy and efficient. The technical design of this invention enabled millions of lawn lovers and gardeners the opportunity to water their plants and vegetables, evenly, without causing damage to the leaves and stems or over-watering them. His invention has been modified

over the years to take on larger tasks such as watering crops and vineyards.

**Inventor:** W. H. Sammons
**Invention:** The Ironing Comb
**Location:** Unknown **Date:** December 21, 1920 **Patent #:** 1,362,823
During the 1920's many African-American women started to take greater pride in their physical beauty. W. H. Sammons invented the comb, also known as the straightening comb, which made it possible for Black women to become more creative in styling their hair in the styles of that period. Although there have been some improvements in the comb it is still used today as it was invented in the 1920's.

**Inventor:** Thomas W. Steward
**Invention:** Mop
**Location:** Detroit, MI **Date:** June 13, 1893 **Patent #:** 499,402
Cleaning floors before the invention of the mop was a very tedious and painful job. The floors were cleaned prior to the introduction of the mop by the use of brushes, scrub pads and rags which often times caused unbearable pain to the joints and muscles of the cleaner. Mr. Steward's invention, the mop, made the task of cleaning floors a lot easier. His invention is still just as popular today among homeowners and businesses as it was when it was invented in the late 1800's.

**Inventor:** George T. Sampson
**Invention:** Clothes Dryer
**Location:** Dayton, OH **Date:** June 7, 1892 **Patent #:** 476,416
The very modern machines that are used today to dry clothes have been improved greatly over the ones used in olden days. Initially, clothing was dried outside on a clothesline. Mr. Sampson invented a portable clothes line that allowed clothing to be dried inside of the house near a furnace or stove.

**Inventor:** William B. Purvis
**Invention:** Hand Stamp
**Location:** Philadelphia, PA **Date:** February 27, 1885 **Patent #:** 273,149

The hand stamp is a practical instrument for labeling packages and boxes that are shipped from stores. The hand stamp is also very convenient for office use. It is made of a metal handle with a rubber pad at the base. The rubber base holds the impression of what is to be stamped. Offices and companies use them to indicate "paid," the name of company, or "past due."

**Inventor:** Sarah Boone
**Invention:** Improvement to the Ironing Board
**Location:** New Haven, CT **Date:** April 26, 1892 **Patent #:** 473,653
An improvement to the ironing board by an African-American woman in 1892 was a very important one. It was especially meaningful for dress makers. Before Sarah Boone's improvement to the front end of the ironing board, it was difficult to iron a man's shirt sleeve and the women's dress sleeve. The narrowing of the front end of the board made the ironing process much easier.

**Inventor:** Charles B. Brooks
**Invention:** Street Sweeper
**Location:** Newark, NJ **Date:** March 17, 1890 **Patent #:** 556,711
It is important for the beautification of a city that its streets are kept clean. Mr. Charles Brooks made an important addition to the initial sweeper by placing brushes on the front end of it that would rotate, pushing glass, dirt and other debris into receptacles. All citizens of a given city can help to keep it clean by placing paper and cans into receptacles.

**Inventor:** George F. Grant
**Invention:** Golf Tee
**Location:** Boston, MA **Date:** December 12, 1899 **Patent #:** 638,920
In the world of sports the game of golf is an intriguing one. The player calls out "fore" as a warning before hitting the ball. The golf ball is hit from a small "T" shaped object raised off the ground. The tee being raised from the ground makes it easier to hit the ball and also protects the turf.

**Inventor:** Lloyd P. Ray
**Invention:** Improvement of the Dust Pan
**Location:** Seattle, WA **Date:** August 3, 1882 **Patent #:** 587,607

An important part of keeping a house clean is the sweeping of its floors. The ways in which particles were removed from floors were awkward before the dust pan was invented. Mr. Ray constructed his invention in such a way that the front edge of the pan was heavier and closer to the floor making it easier to brush dirt into it. This was an improvement over the original dust pan.

## African-American Women's Inventions

**Inventor:** Judy W. Reed
**Invention:** Dough Kneader and Roller
**Location:** Washington, DC **Date:** September 20, 1884 **Patent #**: 305,474
Dough is what we call a mixture of ingredients that make pastries, breads and pizzas. The mixture has more dry ingredients than wet ones and yet keeping the dough moist is important to the success of the final product. To do this well the dough has to be kneaded and rolled. In large quantities this can be hard work. The dough kneader and roller machine prepared large quantities of dough more quickly and with less difficulty. It also had a device that protected the dough from foreign particles.

**Inventor:** Madeline M. Turner
**Invention:** The Fruit Press
**Location:** Oakland, CA **Date:** April 25, 1916 **Patent #**: 1,180,959
Did you ever think about the number of oranges and the work it takes to get one glass of orange juice? Madeline Turner of Oakland, California, did. When she thought about how many oranges it took and how hard one had to work to get a little juice, she invented a fruit press or juicer. Not only is it easier for us to get fruit juice today, but we can get more juice a lot faster by using the fruit press.

**Inventor:** M.V.B. Brown, et al.
**Invention:** Home Security System Utilizing Television Surveillance
**Location:** Jamaica, NY **Date:** December 2, 1969 **Patent #**: 3,482,037
On December 2, 1969, M.V.B. Brown received a patent for a video and audio

security system that could be controlled inside the house. This enabled the occupant to see who was at the door, listen and decide whether or not to let them in.

**Inventor:** Julia T. Hammonds
**Invention:** Apparatus for Holding Yarn Skeins
**Location:** Lebanon, IL **Date:** December 15, 1896 **Patent #:** 572,985
If you have ever knitted, crocheted or woven yarn you know that it is difficult to keep the yarn from tangling while you work. Julia Hammonds devised an apparatus that provided easy access to the yarn while keeping it neatly wound. This invention was designed to fit on a chair so that the knitter could use it while working and then wind leftover yarn back onto the spool when finished.

**Inventor:** Gertrude E. Downing
**Invention:** Corner Cleaning Attachment
**Location:** Washington, DC **Date:** February 13, 1973 **Patent #:** 3,715,772
How happy housekeepers were when the vacuum cleaner was invented. There was just one problem. The machines didn't work in corners. Gertrude E. Downing of Washington, D.C., solved that problem by developing an attachment constructed like a corner to go into corners and pull the dirt up into the machine. This attachment had spinning brushes to scrub the floor loosening the particles and could be attached to industrial floor cleaners that polished and waxed as well.

**Inventor:** Mary J. Reynolds
**Invention:** Hoisting and Loading Mechanism
**Location:** St. Louis, MO **Date:** April 20, 1920 **Patent #:** 1,337,667
On April 20, 1920, Mary Jane Reynolds received a patent for a machine that helped workmen lift heavy loads onto trucks and railroad cars. This machine lifts the freight and loads it, making it faster and easier to ship large loads.

**Inventor:** Alice H. Parker
**Invention:** Improvement on the Heating Furnace

**Location:** Morristown, NJ **Date:** December 23, 1919 **Patent #:** 1,325,905
There was once a time when the temperature of a room depended on the heating device in each room. Then furnaces were developed that allowed more than one room to be heated at one time. This was better, with one exception; since heat went into each room from one burner, smaller rooms were too hot and larger rooms did not get well heated. Alice H. Parker improved the heating furnace by developing a system that uses multiple burners and a numerically controlled thermostat. Each room had a number that was assigned a particular burner. This way one system could heat many rooms at one time or only the rooms that needed it.

**Inventor:** Lydia D. Newman
**Invention:** Improved Model of the Hair Brush
**Location:** New York, NY **Date:** November 15, 1898 **Patent #:** 614,335
Do you often have to clean your hair brush? Have you thought sometimes, "there must be an easier way?" Lydia D. Newman thought of one, designed it and sells it even today. When we brush our hair, we do more than style it. Brushing the hair promotes a healthy scalp and cleaner hair, however, the dirt and hair often get trapped in the brush. Lydia Newman designed a hair brush that has a detachable bristle holder which traps the dirt and hair between the bristles and the handle. Sliding the holder off the handle exposes the residue and makes it easier to clean the hair brush.

**Inventor:** Virgle M. Ammonds
**Invention:** Fireplace Damper Actuating Tool
**Location:** Eglon, WV **Date:** September 30, 1975 **Patent #:** 3,908,633
Before 1975 homes that had fireplaces had a problem when it was not chilly enough for a fire, but it was too chilly for the doors and windows to be open. As long as there was no way to keep the damper tightly closed, cold air could blow into the house from the chimney. We now have cleaner, more efficient fireplaces because Virgle Ammonds invented a device that allowed the fireplace damper to be tightly closed.

**Inventor:** Lydia M. Holmes
**Invention:** Knockdown Wheeled Toy

**Location:** St. Augustine, FL **Date:** November 14, 1950 **Patent #:** 2,529,692
When your parents were children, they played with toys that didn't need electricity or batteries to work. In order for them to move their toys, the toys were constructed with wheels and pull strings. Often, parents assembled the toys with their children. Doing this helped children learn how to build things. Lydia Holmes, an inventor from St. Augustine, Florida, designed the plans for an easily assembled wooden bird, truck or dog.

**Inventor:** Sara E. Goode
**Invention:** The Cabinet Bed
**Location:** Chicago, IL **Date:** July 14, 1883 **Patent #:** 322,177
At your house, are there ever times when your family needs an extra bed but there just isn't anywhere to put it? The cabinet bed, when folded was a solution for people in 1885. It looked like a piece of furniture when the bed part was tucked inside.

**Inventor:** Henrietta Bradberry
**Invention:** The Bed Rack
**Location:** Chicago, IL **Date:** May 25, 1943 **Patent #:** 2,320,027
On May 25, 1943, Mrs. Henrietta Bradberry received patent number 2,320,027 for a device called a Bed Rack. The Bed Rack was attached to the bed and the sheets and blankets were pulled through it so that they could air out before the bed was made. Beds used to be very heavy and much larger so taking the sheets and blankets off to air was a difficult job, especially since you had to put them back on again.

## Food Preparation Made Easier

**Inventor:** R.P. Scott
**Invention:** Corn Silker
**Location:** Unknown **Date:** August 7, 1894 **Patent #:** 524,223
Corn is a very delicious dish and can be cooked in many different ways; on-the-cob, stewed or fried. It has an outer green husk and fibers on the inside called silk. The silk fibers on the inside of corn can be difficult to remove, and

the removal of the silk takes a lot of time. The corn silker is a gadget which speeds up the removal of the silks.

**Inventor:** F.J. Wood
**Invention:** Potato Digger
**Location:** Unknown **Date:** April 23, 1895 **Patent #:** 537,953
Harvesting potatoes was a slow and difficult process before the invention of the potato digger. The digger is pushed into the ground to soften the earth around the potato, making it easier to be scooped up by the picker. This gadget is a very important tool to the farmer.

**Inventor:** Albert Richardson
**Invention:** Improvement of the Butter Churn
**Location:** South Frankfort, MI **Date:** February 17, 1891 **Patent #:** 446,470
Many years ago, on farms in America, farmers used a wooden barrel-like appliance called a churn to make butter. The butter was made from the cream skimmed from the milk which came from the cow. The churn had a long wooden stick coming through a hole in the center of the top. The method of pushing the stick down and pulling it up again separated or skimmed the cream from the milk. Mr. Richardson improved the churn in two ways. He placed glass on two sides of the churn making it easier to tell when the butter was ready without opening it. He also placed a plate inside the churn making it easier to remove the butter.

**Inventor:** J.T. White
**Invention:** Lemon Squeezer
**Location:** Unknown **Date:** December 8, 1896 **Patent #:** 572,849
There are many types of juicers to squeeze juice from various fruits. The lemon squeezer was the first such juicer. This invention made it less difficult to get all of the juice out of the lemon without the seeds and a lot of pulp. This gadget also is neat because it stopped the juice from squirting.

**Inventor:** Norbert Rillieux
**Invention:** Improvement of the Sugar Making Process
**Location:** New Orleans, LA **Date:** December 10, 1846 **Patent #:** 4879

Norbert Rillieux, a native of the state of Louisiana, invented a new and improved technique of processing sugar which involved heating, evaporating and cooling of the liquids intended for the manufacture of sugar. His invention was the continuation of a process for refining sugar which is now very modern and sophisticated.

**Inventor:** Willis Johnson
**Invention:** Egg Beater
**Location:** Cincinnati, OH **Date:** February 5, 1884 **Patent #**: 292,821
The creation of the egg beater in the late 1800's simplified many tasks performed in the kitchen. Mr. Johnson was very thoughtful in making a machine with double-acting blades. While one kind of batter was being mixed on one side to make candy or cookies the other side could blend another kind of mixture. This invention made it possible not only to blend ingredients better, but also faster.

**Inventor:** John Stanard
**Invention:** Improvement of the Icebox
**Location:** Newark, NJ **Date:** July 14, 1891 **Patent #**: 455,891
In years past, the method of preserving leftover food within the household was through the medium of an icebox. The ice for the icebox was bought from an iceman who sold it in blocks of 50 to 100 pounds to households within the neighborhood. John Stanard added two features to the original icebox which improved it greatly. The airducts or holes in special areas of it helped the air to circulate more evenly keeping the food fresher. He also provided a separate compartment in the icebox for drinking water. This feature helped to keep the water from picking up odors from the food.

**Inventor:** A.E. Long and A.A. Jones
**Invention:** Cap for Bottles and Jars
**Location:** Unknown **Date:** September 13, 1898 **Patent #**: 610,715
The invention of the cap for soda bottles and other type jars made travel with food and drink much easier. After this creative invention, soft drinks and food items could be kept fresh and clean, and travel was made more convenient and enjoyable.

**Inventor:** George W. Kelley
**Invention:** Steam Table
**Location:** Norfolk, VA **Date:** October 26, 1897 **Patent #:** 592,591
The creation of the steam table allows food to be cooked early in the day and kept warm or hot to be eaten later. This piece of equipment is used in schools, cafeterias, restaurants and hospitals. Underneath the steam table boiling water is circulated through pipes which provide the heat for the food kept hot on top of it. The portable steam table can be moved from table to table in an establishment for different food items.

**Inventor:** Alexander P. Ashbourne
**Invention:** Biscuit Cutter
**Location:** Oakland, CA **Date:** November 30, 1875 **Patent #:** 170,460
Biscuits as well as cookies can be cut and shaped in different sizes with the creation of a cutter in 1875 by a native of the state of California. Biscuit or cookie cutters have plates and springs inside so that when pressed down into dough, the desired shape or design chosen is cut. Today, one can find a simplified design of cutters sold in all sizes and shapes with designs for special days including Christmas and Halloween.

**Inventor:** Alfred L. Cralle
**Invention**: Ice Cream Mold and Disher
**Location:** Pittsburgh, PA **Date:** February 2, 1897 **Patent #:** 576,395
Eating ice cream is very enjoyable and is what can be called an American favorite. Many children like to eat it from a cone. The ice cream is put into the cone with a disher, known today as a scoop. Alfred Cralle, all people of America thank you for the round portions of ice cream that can now be scooped from a larger container, easily and conveniently.

## Innovations of the Old West

**Inventor:** Jan E. Matzeliger
**Invention:** Shoe Lasting Machine

**Location:** Unknown **Date:** March 20, 1883 **Patent #:** 274,207
In early days of the United States, many people traveled West on foot in search of gold. They walked many miles across mountains and plains. This means of travel on foot meant that there was a constant need for shoes and boots. The shoe lasting machine made shoe repair faster. The shoe cobbler could sew the tops of shoes and boots to the soles in a more efficient way. With this machine being used to make repairs to shoes they would also last longer.

**Inventor:** William D. Davis
**Invention:** Riding Saddle
**Location:** Fort Assinniboine, MT **Date:** October 6, 1896 **Patent #:** 568,939
In America, horse racing is a favorite pastime. The Kentucky Derby is one of the most famous races run annually. The riding saddle makes the jockey comfortable as he rides his horse to the finish line. The riding saddle also makes the jockey's ride safer. Mr. Davis' invention improved the saddle by making it easier to adjust, thus adding elasticity to the seat of the saddle. He also improved durability of the stirrup, saddle, strings and straps.

**Inventor:** M.C. Hamey
**Invention:** Lantern or Lamp
**Location:** Unknown **Date:** August 19, 1884 **Patent #:** 303,844
The use of electricity and the light bulb are modern day means of lighting our homes and streets. Before the era of the light bulb, people used lanterns to light the streets, their homes, carriages, and also to send signals. In olden days, one single lantern could be seen miles away on very dark nights. Men who worked on the railroad used the lantern to communicate with each other and to warm the train's conductors.

**Inventor:** M.C. Hamey
**Invention:** Overboot for Horses
**Location:** Washington, DC **Date:** April 19, 1892 **Patent #:** 473,295
During rain and sleet, a horse has difficulty keeping its balance. The overboot was invented to keep the horse from slipping and sliding. The legs of the horse are vital to its body and also the most fragile part of it. The horse could be seriously injured if it should lose its footing and fall.

**Inventor:** George Cooke
**Invention:** Automatic Fishing Device
**Location:** Louisville, KY **Date:** May 30, 1899 **Patent #:** 625,829
The reel is a very important gadget in helping the fisherman catch fish. Once a fish bites the bait of the person fishing, it will strive to swim away. This automatic device helps to take up the "slack" in the reel and line, allowing the fish to be reeled in by the fisherman.

**Inventor:** John W. West
**Invention:** Running Gear
**Location:** Saylorville, IA **Date:** October 8, 1870 **Patent #:** 108,419
With the creation of the running gear, travel by wagon became safe and more efficient. The inventor, John West, by making the hind wheels larger than the front wheels and the hind axle more elevated than its front axle, helped to balance the weight of the load in such a manner so as to help propel and push the load forward. This invention made it less strenuous on the horse and travel by wagon was made faster.

**Inventor:** C.O. Bailiff
**Invention:** Shampoo Head Rest
**Location:** Kalamazoo, MI **Date:** October 11, 1898 **Patent #:** 612,008
Women take a great deal of pride in keeping themselves well groomed. In the old West, if a woman wore her hair long, it was especially hard for her to wash it while bending over a tub. The invention of a headrest made shampooing more comfortable. It was attached to the back of the chair giving support to the head so that the person could lean backwards allowing the water to run down the hair instead of over the face.

**Inventor:** Henry Brown
**Invention:** Desktop File Box
**Location:** Washington, DC **Date:** November 2, 1886 **Patent #:** 352,036
It is important to keep the table or desk where various tasks are performed, orderly. In school or the work place, one's desk can become cluttered with pens, pencils, paper clips, writing papers, etc. The desktop file box lended itself primarily for filing away carbon paper. The invention of the file box helped to keep carbon paper neat and the white paper not being used clean.

# Appendix II

## Other African-American Inventions

### Patent Index

| Inventor | Invention | Date | Patent # |
|---|---|---|---|
| Abrams, W.B. | Hame Attachment | Apr. 14, 1891 | 450,550 |
| Allen, C.W. | Self-Leveling Table | Nov. 1, 1898 | 613,436 |
| Allen, J.B. | Clothes Line Support | Dec. 10, 1895 | 551,105 |
| Ashbourne, A.P. | Process/Preparing Coconut for Domestic Use | June 1, 1875 | 163,962 |
| Ashbourne, A.P. | Biscuit Cutter | Nov. 30, 1875 | 170,460 |
| Ashbourne, A.P. | Refining Coconut Oil | July 27, 1880 | 230,518 |
| Ashbourne, A.P. | Process of Treating Coconut | Aug. 21, 1877 | 194,287 |
| Bailes, Wm. | Ladder Scaffold-Support | Aug. 5, 1879 | 218,154 |
| Bailey, L.C. | Combined Truss and Bandage | Sept. 25, 1883 | 285,545 |
| Bailey, L.C. | Folding Bed | July 18, 1899 | 629,286 |
| Bailiff, C.O. | Shampoo Headrest | Oct. 11, 1898 | 12,008 |
| Ballow, W.J. | Combined Hatrack and Table | Mar. 29, 1898 | 601,422 |

| | | | |
|---|---|---|---|
| Barnes, G.A.E. | Design for Sign | Aug. 19, 1898 | 29,193 |
| Becket, G.E. | Letter Box | Oct. 4, 1892 | 483,325 |
| Bell, L. | Locomotive Smoke Stack | May 23, 1871 | 115,153 |
| Bell, L. | Dough Kneader | Dec. 10, 1872 | 133,823 |
| Benjamin, L.W. | Broom Moisteners and Bridles | May 16, 1893 | 497,747 |
| Benjamin, M.E. | Gong and Signal Chairs for Hotels | July 17, 1888 | 386,286 |
| Binga, M.W. | Street Sprinkling Apparatus | July 22, 1879 | 217,843 |
| Blackburn, A.B. | Railway Signal | Jan. 10, 1888 | 376,362 |
| Blackburn, A.B. | Spring Seat for Chairs | Apr. 03, 1888 | 380,420 |
| Blackburn, A.B. | Cash Carrier | Oct. 23, 1888 | 391,577 |
| Blair, Henry | Corn Planter | Oct. 14, 1834 | — |
| Blair, Henry | Cotton Planter | Aug. 31, 1836 | — |
| Blue, L. | Hand Corn Shelling Device | May 20, 1884 | 298,937 |
| Booker, L.F. | Design Rubber Scraping Knife | Mar. 28, 1899 | 30,404 |
| Boone, Sarah | Ironing Board | Apr. 26, 1892 | 473,653 |
| Bowman, H.A. | Making Flags | Feb. 23, 1892 | 469,395 |
| Brooks, C.B. | Punch | Oct. 31, 1893 | 507,672 |

| | | | |
|---|---|---|---|
| Brooks, C.B. | Street Sweepers | Mar. 17, 1896 | 556,711 |
| Brooks, C.B. | Street Sweepers | May 12, 1896 | 560,154 |
| Brooks, Hallstead & Page | Street Sweepers | Apr. 21, 1896 | 558,719 |
| Brown, Henry | Receptacle for Storing & Preserving Paper | Nov. 02, 1886 | 352,036 |
| Brown, L.F. | Bridle Bit | Oct. 25, 1892 | 484,994 |
| Brown, O.E. | Horseshoe | Aug. 23, 1892 | 481,271 |
| Brown & Latimer | Water Closets for Railway Cars | Feb. 10, 1874 | 147,363 |
| Burr, J.A. | Lawn Mower | May 09, 1899 | 624,749 |
| Burr, W.F. | Switching Device for Railways | Oct. 31, 1899 | 636,197 |
| Burwell, W. | Boot or Shoe | Nov. 28, 1899 | 638,143 |
| Butler, R.A. | Train Alarm | June 15, 1897 | 584,540 |
| Butts, J.W. | Luggage Carrier | Oct. 10, 1899 | 634,611 |
| Byrd, T.J. | Improvement in Holders of Reins for Horses | Feb. 06, 1872 | 123,328 |
| Byrd, T.J. | Apparatus for Detaching Horses from Carriages | Mar. 19, 1872 | 124,790 |
| Byrd, T.J. | Improvement in Neck Yokes for Wagons | Apr. 30, 1872 | 126,181 |
| Byrd, T.J. | Improvement in Car-Couplings | Dec. 01, 1874 | 157,370 |

| | | | |
|---|---|---|---|
| Campbell, W.S. | Self-Setting Animal Trap | Aug. 30, 1881 | 246,369 |
| Cargill, B.F. | Invalid Cot | July 25, 1899 | 629,658 |
| Carrington, T.A. | Range | July 25, 1876 | 180,323 |
| Carter, W.C. | Umbrella Stand | Aug. 04, 1885 | 323,397 |
| Certain, J.M. | Parcel Carrier for Bicycles | Dec. 26, 1899 | 639,708 |
| Cherry, M.A. | Velocipede | May 08, 1888 | 382,351 |
| Cherry, M.A. | Street Car Fender | Jan. 01, 1895 | 531,908 |
| Church, T.S. | Carpet Beating Machine | July 29, 1884 | 302,237 |
| Clare, O.B. | Trestle | Oct. 09, 1888 | 390,753 |
| Coates, R. | Overboot for Horses | Apr. 19, 1892 | 473,295 |
| Cook, G. | Automatic Fishing Device | May 30, 1899 | 625,829 |
| Coolidge, J.S. | Harness Attachment | Nov. 13, 1888 | 392,908 |
| Cooper, J. | Shutter & Fastening | May 01, 1883 | 276,563 |
| Cooper, J. | Elevator Device | Apr. 02, 1895 | 536,605 |
| Cooper, J. | Elevator Device | Sept. 21, 1897 | 590,257 |
| Cornwell, P.W. | Draft Regulator | Oct. 02,1888 | 390,284 |
| Cornwell, P.W. | Draft Regulator | Feb. 07, 1893 | 491,082 |
| Cralle, A.L. | Ice-Cream Mold | Feb. 02, 1897 | 576,395 |

| | | | |
|---|---|---|---|
| Creamer, H. | Steam Feed Water Trap | Mar. 17, 1895 | 313,854 |
| Creamer, H. | Steam Trap Feeder | Dec. 11, 1888 | 94,463 |
| (Creamer also patented five steam traps between 1887 and 1893) | | | |
| Cosgrove, W.F. | Automatic Stop Plug for Gas Oil Pipes | Mar. 17, 1885 | 313,993 |
| Darkins, J.T. | Ventilation Aid | Feb. 19, 1895 | 534,322 |
| Davis, I.D. | Tonic | Nov. 02, 1886 | 351,829 |
| Davis, W.D. | Riding Saddles | Oct. 06, 1896 | 568,939 |
| Davis, W.R., Jr. | Library Table | Sept. 24, 1878 | 208,378 |
| Deitz, W.A. | Shoe | Apr. 30, 1867 | 64,205 |
| Dickinson, J.H. | Pianola | Detroit, MI, 1899 | n/a |
| Dorsey, O. | Door-Holding Device | Dec. 10, 1878 | 210,764 |
| Dorticus, C.J. | Device/Applying Coloring/ Sides of Shoe Soles | Mar. 19, 1895 | 535,820 |
| Dorticus, C.J. | Machine for Embossing Photo | Apr. 16, 1895 | 537,442 |
| Dorticus, C.J. | Photographic Print Wash | Apr. 23, 1895 | 537,968 |
| Dorticus, C.J. | Hose Leak Stop | July 18, 1899 | 629,315 |
| Downing, P.B. | Electric Switch for Railroad | June 17, 1890 | 430,118 |
| Downing, P.B. | Letter Box | Oct. 27, 1891 | 462,093 |
| Downing, P.B. | Street Letter Box | Oct. 27, 1891 | 462,093 |
| Dunnington, J.H. | Horse Detachers | Mar. 16, 1897 | 578,979 |

| | | | |
|---|---|---|---|
| Edmonds, T.H. | Separating Screens | July 20, 1897 | 586,724 |
| Elkins, T. | Dining/Ironing Table/ Quilting Frame Combined | Feb. 22, 1870 | 100,020 |
| Elkins, T. | Chamber Commode | Jan. 09, 1872 | 122,518 |
| Elkins, T. | Refrigerating Apparatus | Nov. 04,1879 | 221,222 |
| Evans, J.H. | Convertible Settees | Oct. 05, 1897 | 591,095 |
| Faulkner, H. | Ventilated Shoe | Apr. 29,1890 | 426,495 |
| Ferrell, F.J. | Steam Trap | Feb. 11, 1890 | 420,993 |
| Ferrell, F.J. | Apparatus for Melting Snow | May 27,1890 | 428,670 |

(Ferrell also patented eight valves between 1890 and 1893.)

| | | | |
|---|---|---|---|
| Fisher, D.A. | Joiner's Clamp | Apr. 20, 1875 | 162,281 |
| Fisher, D.A. | Furniture Castor | Mar. 14, 1876 | 174,794 |
| Forten, J. | Sail Control | Mass. Newspaper, 1850 | |
| Goode, Sarah E. | Folding Cabinet Bed | July 14, 1885 | 322,177 |
| Grant, W.S. | Curtain Rod Support | Aug. 04, 1896 | 565,075 |
| Gray, R.H. | Baling Press | Aug. 28, 1894 | 525,203 |
| Gray, R.H. | Cistern Cleaners | Apr. 09, 1895 | 537,151 |
| Gregory, J. | Motor | Apr. 26, 1887 | 361,937 |
| Grenon, H. | Razor Stropping Device | Feb. 18, 1896 | 554,867 |

| | | | |
|---|---|---|---|
| Griffing, F.W. | Pool Table Attachment | June 13, 1899 | 626,902 |
| Gunn, S.W. | Boot or Shoe | Jan. 16, 1900 | 641,642 |
| Haines, J.H. | Portable Basin | Sept. 28, 1897 | 590,833 |
| Hammonds, J.F. | Apparatus for Holding Yarn Skeins | Dec. 15, 1886 | 572,985 |
| Harding, F.H. | Extension Banquet Table | Nov. 22, 1898 | 614,468 |
| Hawkins, J. | Gridiron | Mar. 26, 1845 | 3,973 |
| Hawkins, R. | Harness Attachment | Oct. 04, 1887 | 370,943 |
| Headen, M. | Foot Power Hammer | Oct. 05, 1886 | 350,363 |
| Hearness, R. | Sealing Attachment for Bottles | Feb. 15, 1898 | 598,929 |
| Hearness, R. | Detachable Car Fender | July 04, 1899 | 628,003 |
| Hilyer, A.F. | Water Evaporator Attach Hot Air Registers | Aug. 26, 1890 | 435,095 |
| Hilyer, A.F. | Registers | Oct. 14, 1890 | 438,159 |
| Holmes, E.H. | Gage | Nov. 12, 1895 | 549,513 |
| Hunter, J.H. | Portable Weighing Scales | Nov. 03, 1896 | 570,553 |
| Hyde, R.N. | Composition for Cleaning/ Preserving Carpets | Nov. 06, 1888 | 392,205 |

| | | | |
|---|---|---|---|
| Jackson, B.F. | Heating Apparatus | Mar. 01, 1898 | 599,985 |
| Jackson, B.F. | Matrix Drying Apparatus | May 10, 1898 | 603,879 |
| Jackson, B.F. | Gas Burner | Apr. 04, 1899 | 622,482 |
| Jackson, H.A. | Kitchen Table | Oct. 06, 1896 | 596,135 |
| Jackson, W.H. | Railway Switch | Mar. 09, 1897 | 578,641 |
| Jackson, W.H. | Railway Switch | Mar. 16, 1897 | 593,665 |
| Jackson, W.H. | Automatic Locking Switch | Aug. 23, 1898 | 609,436 |
| Johnson, D. | Rotary Dining Table | Jan. 15, 1888 | 396,089 |
| Johnson, D. | Lawn Mower Attachment | Sept. 10, 1889 | 410,836 |
| Johnson, D. | Grass Receivers for Lawn Mowers | June 10, 1890 | 429,629 |
| Johnson, I.R. | Bicycle Frame | Oct. 10, 1899 | 634,823 |
| Johnson, P. | Swinging Chairs | Nov. 15, 1881 | 249,530 |
| Johnson, P. | Eye Protector | Nov. 02, 1880 | 234,039 |
| Johnson, W. | Velocipede | June 20, 1899 | 627,335 |
| Johnson, W.A. | Paint Vehicle | Dec. 04,1888 | 393,763 |
| Johnson, W.H. | Overcoming Dead Centers | Feb. 04, 1896 | 554,223 |
| Johnson, W.H. | Overcoming Dead Centers | Oct. 11, 1898 | 612,345 |
| Johnson, W. | Egg Beater | Feb. 05, 1884 | 292,821 |

| | | | |
|---|---|---|---|
| Jones, F.M. | Ticket Dispensing Machine | June 27, 1939 | 2,163,754 |
| Jones, F.M. | Air Conditioning Unit | July 12, 1949 | 2,475,841 |
| Jones, F.M. | Method for Air Conditioning | Dec. 0 7, 1954 | 2,696,086 |
| Jones, F.M. | Method for Preserving Perishables | Feb. 12, 1957 | 2,780,923 |
| Jones, F.M. | Two-Cycle Gasoline Engine | Nov. 28, 1950 | 2,523,273 |
| Jones, F.M. | Two-Cycle Gas Engine | May 29, 1945 | 2,376,968 |
| Jones, F.M. | Starter Generator | July 12, 1949 | 2,475,842 |
| Jones, F.M. | Starter Generator for Cooling Gas Engines | n/a | 2,475,843 |
| Jones, F.M. | Two-Cycle Gas Engine | Mar. 11, 1947 | 2,417,253 |
| Jones, F.M. | Means/Thermostatically Operating Gas Engines | July 26, 1949 | 2,477,377 |
| Jones, F.M. | Rotary Compressor | Apr. 18,1950 | 2,504,841 |
| Jones, F.M. | System/Controlling Refrigeration Units | May 23, 1950 | 2,509,099 |
| Jones, F.M. | Heating/Cooling Atmosphere within Enclosure | Oct. 24, 1950 | 2,526,874 |
| Jones, F.M. | Prefabricated Refrigerator Construction | Dec. 26, 1950 | 2,535,682 |
| Jones, F.M. | Refrigeration Control Device | Jan. 08, 1952 | 2,581,956 |
| Jones, F.M. | Methods, Means of Defrosting a Cold Diffuser | Jan. 19, 1954 | 2,666,298 |
| Jones, F.M. | Control Device for Internal Combustion Engine | Sept. 02, 1958 | 2,850,001 |
| Jones, F.M. | Thermostat and Temperature Control System | Feb. 23, 1960 | 2,926,005 |
| Jones, F.M. | Means for Automatically Stopping & Starting Gas Engines ("J.A. Numero, et al") | Dec. 21, 1943 | 2,337,164 |

| | | | |
|---|---|---|---|
| Jones, F.M. | Design for Air Conditioning Unit | July 4, 1950 | 159,209 |
| Jones, F.M. | Design for Air Conditioning Unit | Apr. 28, 1942 | 132,182 |
| Jones & Long | Caps for Bottles | Sept. 13, 1898 | 610,715 |
| Joyce, J.A. | Ore Bucket | Apr. 26,1898 | 603,143 |
| Latimer & Brown | Water Closets for Railway Cars | Feb. 10, 1874 | 147,363 |
| Latimer & Nichols | Electric Lamp | Sept. 13, 1881 | 247,097 |
| Latimer & Tregoning | Globe Support for Electric Lamps | Mar. 21, 1882 | 255,212 |
| Lee, H. | Animal Trap | Feb. 12, 1867 | 61,941 |
| Lee, J. | Kneading Machine | Aug. 07, 1894 | 524,042 |
| Lee, J. | Bread Crumbing Machine | June 04, 1895 | 540,553 |
| Leslie, F.W. | Envelope Seal | Sept. 21, 1897 | 590,325 |
| Lewis, A.L. | Window Cleaner | Sept. 27, 1892 | 483,359 |
| Lewis, E.R. | Spring Gun | May 03, 1887 | 362,096 |
| Linden, H. | Piano Truck | Sept. 08, 1891 | 459,365 |
| Little, E. | Bridle-Bit | Mar. 0 7, 1882 | 254,666 |
| Loudin, F.J. | Sash Fastener | Dec. 12, 1892 | 510,432 |
| Loudin, F.J. | Key Fastener | Jan. 09, 1894 | 512,308 |

| | | | |
|---|---|---|---|
| Love, J.L. | Plasterers' Hawk | July 09, 1895 | 542,419 |
| Love, J.L. | Pencil Sharpener | Nov. 23, 1897 | 594,114 |
| Marshall, T.J. | Fire Extinguisher | May 26, 1872 | 125,063 |
| Marshall, W. | Grain Binde | May 11, 1886 | 341,599 |
| Martin, W.A. | Lock | July 23, 1889 | 407,738 |
| Martin, W.A. | Lock | Dec. 30, 1890 | 443,945 |
| Matzeliger, J.E. | Mechanism for Distributing Tacks | Nov. 26, 1899 | 415,726 |
| Matzeliger, J.E. | Nailing Machine | Feb. 25, 1896 | 421,954 |
| Matzeliger, J.E. | Tack Separating Mechanism | Mar. 25, 1890 | 423,937 |
| McCoy, E.J. | Lubricator | Mar. 28, 1882 | 255,443 |
| McCoy, E.J. | Lubricator | July 18,1882 | 261,166 |
| McCoy, E.J. | Lubricator | Feb. 08, 1887 | 357,491 |
| McCoy, E.J. | Lubricator | May 29, 1888 | 383,745 |
| McCoy, E.J. | Lubricator | May 29, 1888 | 383, 746 |
| McCoy, E.J. | Lubricator | Dec. 24, 1899 | 418,139 |
| McCoy, E.J. | Lubricator | Dec. 29, 1891 | 465,875 |
| McCoy, E.J. | Lubricator | Apr. 05, 1892 | 472,066 |
| McCoy, E.J. | Lubricator | Sept. 13, 1898 | 610,634 |
| McCoy, E.J. | Lubricator | Oct. 04, 1898 | 611,759 |
| McCoy, E.J. | Oil Cup | Nov. 15, 1898 | 614,307 |
| McCoy, E.J. | Lubricator | June 27, 1899 | 627,623 |
| McCoy, E.J. | Lubricator for Steam Engines | July 02, 1872 | 129,843 |
| McCoy, E.J. | Lubricator for Steam Engines | Aug. 06, 1872 | 130,305 |
| McCoy, E.J. | Steam Lubricator | Jan. 20, 1874 | 146,697 |
| McCoy, E.J. | Ironing Table | May 12, 1874 | 150,876 |

| | | | |
|---|---|---|---|
| McCoy, E.J. | Steam Cylinder Lubricator | Feb. 01, 1876 | 173,032 |
| McCoy, E.J. | Steam Cylinder Lubricator | July 04, 1876 | 179,585 |
| McCoy, E.J. | Steam Dome | June 16, 1885 | 320,354 |
| McCoy, E.J. | Lubricator Attachment | Apr. 19, 1887 | 361,435 |
| McCoy, E.J. | Lubricator for Safety Valves | May 24, 1887 | 363,529 |
| McCoy, E.J. | Drip Cup | Sept. 29, 1891 | 460,215 |
| McCoy & Hodges | Lubricator | Dec. 24, 1889 | 418,139 |
| McCree, D. | Portable Fire Escape | Nov. 11, 1890 | 440,322 |
| Mendenhall, A. | Holder for Driving Reins | Nov. 28, 1899 | 637,811 |
| Miles, A. | Elevator | Oct. 11, 1887 | 371,207 |
| Mitchell, J.M. | Cheek Row Corn Planter | Jan. 16, 1900 | 641,462 |
| Moody, W.U. | Game Board Design | May 11, 1897 | 27,046 |
| Morehead, K. | Reel Carrier | Oct. 06, 1896 | 568,916 |
| Murray, G.W. | Combined Furrow Opener and Stalk-Knocker | Apr. 10, 1894 | 517,960 |
| Murray, G.W. | Cultivator and Marker | Apr. 10, 1894 | 517,961 |
| Murray, G.W. | Planter | June 05, 1894 | 520,887 |
| Murray, G.W. | Cotton Chopper | June 05, 1894 | 520,888 |
| Murray, G.W. | Fertilizer Distributor | June 05, 1894 | 520,889 |
| Murray, G.W. | Planter | June 05, 1894 | 520,890 |
| Murray, G.W. | Combined Cotton Seed | June 05, 1894 | 520,891 |

| | | | |
|---|---|---|---|
| Murray, G.W. | Planter and Fertilizer Distributor Reaper | June 05, 1894 | 520,892 |
| Murray, G.W. | Attachment for Bicycles | Jan. 27, 1891 | 445,452 |
| Nance, L. | Game Apparatus | Dec. 01, 1891 | 464,035 |
| Nash, H.H. | Life Preserving Stool | Oct. 05, 1875 | 168,519 |
| Newman, L.D. | Brush | Nov. 15, 1898 | 614,335 |
| Newson, S. | Oil Heater or Cooker | May 22, 1894 | 520,188 |
| Nichols & Latimer | Electric Lamp | Sept. 13, 1881 | 247,097 |
| Nickerson, W.J. | Mandolin and Guitar Attachment for Pianos | June 27, 1899 | 627,739 |
| O'Connor & Turner | Alarm for Boilers | Aug. 25, 1896 | 566,612 |
| O'Connor & Turner | Steam Gage | Aug. 25, 1896 | 566,613 |
| O'Connor & Turner | Alarm for Coasts Containing Vessel | Feb. 08, 1898 | 598,572 |
| Outlaw, J.W. | Horseshoes | Nov. 15, 1898 | 614,273 |
| Perryman, F.R. | Caterers' Tray Table | Feb. 02, 1892 | 468,038 |
| Peterson, H. | Attachment for Lawn Mowers | Apr. 30, 1889 | 402,189 |
| Phelps, W.H. | Apparatus for Washing Vehicles | Mar. 23, 1897 | 579,242 |
| Pickering J.F. | Air Ship | Feb. 20, 1900 | 643,975 |
| Pickett, H. | Scaffold | June 30, 1874 | 152,511 |

| | | | |
|---|---|---|---|
| Pinn, T.B. | File Holder | Aug. 17, 1880 | 231,355 |
| Polk, A.J. | Bicycle Support | Apr. 14, 1896 | 558,103 |
| Pugsley, A. | Blind Stop | July 29, 1890 | 433,306 |
| Purdy & Peters | Design for Spoons | Apr. 23, 1895 | 24,228 |
| Purdy & Sadgwar | Folding Chair | June 11, 1889 | 405,117 |
| Purdy, W. | Device for Sharpening Edged Tools | Oct. 27, 1896 | 570,337 |
| Purdy, W. | Device for Sharpening Edged Tools | Aug. 16, 1898 | 609,367 |
| Purdy, W. | Device for Sharpening Edged Tools | Aug. 01, 1899 | 630,106 |
| Purvis, W.B. | Bag Fastener | Apr. 25, 1882 | 256,856 |
| Purvis, W.B. | Hand Stamp | Feb. 27, 1883 | 273,149 |
| Purvis, W.B. | Fountain Pen | Jan. 07, 1890 | 419,065 |
| Purvis, W.B. | Electric Railway | May 01, 1894 | 519,291 |
| Purvis, W.B. | Magnetic Car Balancing Device | May 21, 1895 | 539,542 |
| Purvis, W.B. | Electric Railway Switch | Aug. 17, 1897 | 588,176 |

(Purvis also patented ten paper bag machines between 1884 & 1894)

| | | | |
|---|---|---|---|
| Queen, W. | Guard for Companion Ways and Hatches | Aug. 18, 1891 | 458,131 |
| Ray, E.P. | Chair Supporting Device | Feb. 21, 1899 | 620,078 |
| Ray, L.P. | Dust Pan | Aug. 03, 1897 | 587,607 |

| | | | |
|---|---|---|---|
| Reed, J.W. | Dough Kneader and Roller | Sept. 23, 1884 | 305,474 |
| Reynolds, H.H. | Window Ventilator for Railroad Cars | Apr. 03, 1883 | 275,271 |
| Reynolds, H.H. | Safety Gate for Bridges | Oct.0 7, 1890 | 437,937 |
| Reynolds, R.R. | Non-Refillable Bottle | May 02, 1899 | 624,094 |
| Rhodes, J.B. | Water Closets | Dec. 19, 1899 | 639,290 |
| Richardson, A.C. | Hame Fastener | Mar. 14, 1882 | 255,022 |
| Richardson, A.C. | Churn | Feb. 17, 1891 | 446,470 |
| Richardson, A.C. | Casket Lowering Device | Nov. 13, 1894 | 529,311 |
| Richardson, A.C. | Insect Destroyer | Feb. 28, 1899 | 620,362 |
| Richardson, A.C. | Bottle | Dec. 12, 1899 | 638,811 |
| Richardson, W.H. | Cotton Chopper | June 01, 1886 | 343,140 |
| Richardson, W.H. | Child's Carriage | June 18, 1899 | 405,599 |
| Richardson, W.H. | Child's Carriage | June 18,1899 | 405,600 |
| Richey, C.V. | Car Coupling | June 15, 1897 | 584,650 |
| Richey, C.V. | Railroad Switch | Aug. 03, 1897 | 587,657 |
| Richey, C.V. | Railroad Switch | Oct. 26, 1897 | 592,448 |
| Richey, C.V. | Fire Escape Bracket | Dec. 28, 1897 | 596,427 |
| Richey, C.V. | Combined Hammock and Stretcher | Dec. 13, 1898 | 615,907 |
| Rickman, A.L. | Overshoe | Feb. 08, 1898 | 598,816 |
| Ricks, J. | Horseshoe | Mar. 30, 1886 | 338,781 |
| Ricks, J. | Overshoes for Horses | June 6, 1899 | 626,245 |
| Rillieux, N. | Sugar Refiner (Evaporating pan) | Dec. 10, 1846 | 4,879 |

| | | | |
|---|---|---|---|
| Robinson, E.R. | Casting Composite | Nov. 23, 1897 | 594,286 |
| Robinson, E.R. | Electric Railway Trolley | Sept. 19,1893 | 505,370 |
| Robinson, J.H. | Life Saving Guards for Locomotives | Mar. 14, 1899 | 621,143 |
| Robinson, J. | Dinner Pail | Feb. 1, 1887 | 356,852 |
| Romaine, A. | Passenger Register | Apr. 23, 1899 | 402,035 |
| Ross, A.L. | Runner for Stops | Aug. 4, 1896 | 565,301 |
| Ross, A.L. | Bag Closure | June 7, 1898 | 605,343 |
| Ross, A.L. | Trousers Support | Nov. 28, 1899 | 632,549 |
| Roster, D.N. | Bailing Press | Sept. 5, 1899 | 632,539 |
| Ruffin, S. | Vessels for Liquids and Manner of Sealing | Nov. 20, 1899 | 737,603 |
| Russell, L.A. | Guard Attachment for Beds | Aug. 13, 1895 | 544,381 |
| Sampson, G.T. | Sled Propeller | Feb. 17, 1885 | 312,388 |
| Sampson, G.T. | Clothes Drier | June 7, 1892 | 476,416 |
| Scottron, S.R. | Cornice | Jan. 16, 1883 | 270,851 |
| Scottron, S.R. | Pole Tip | Sept. 31, 1886 | 349,525 |
| Scottron, S.R. | Curtain Rod | Aug. 30, 1892 | 481,720 |
| Scottron, S.R. | Supporting Bracket | Sept. 12, 1893 | 505,008 |
| Shanks, S.C. | Sleeping Car Berth Register | July 21, 1897 | 587,165 |
| Shewcraft, F. | Letter Box | n/a | |
| Shorter, D.W. | Feed Rack | May 17, 1887 | 363,089 |

| | | | |
|---|---|---|---|
| Smith, J.W. | Improvement in Games | Apr. 17, 1990 | 647,887 |
| Smith, J.W. | Lawn Sprinkler | Mar. 22, 1898 | 601,065 |
| Smith, J.W. | Lawn Sprinkler | May 4, 1897 | 581,785 |
| Smith, J.D. | Potato Digger | Jan. 1, 1891 | 445,206 |
| Smith, P.D. | Grain Binder | Feb. 23, 1892 | 469,279 |
| Snow & Johns | Liniment | Oct. 7, 1890 | 110,599 |
| Spikes, R.B. | Combination Milk Bottle Opener/ Bottle Cover | June 29, 1926 | 1,590,557 |
| Spikes, R.B. | Method & Apparatus for Obtaining Average Samples & Temperature of Tank Liquids | Oct. 27, 1931 | 1,828,753 |
| Spikes, R.B. | Automatic Gear Shift | Dec. 6, 1932 | 1,828,753 |
| Spikes, R.B. | Transmission & Shifting Thereof | Nov. 28, 1933 | 1,936,996 |
| Spikes, R.B | Self-Locking Rack for Billiard Cues | Around 1910 | Not Found |
| Spikes, R.B. | Automatic Shoe Shine Chair | Around 1939 | Not Found |
| Spikes, R.B | Multiple Barrel Machine Gun | Around 1940 | Not Found |

(Some patents are not included because of current litigation; or because they were so basic in nature that redesigning and refiling procedures are now in process.)

| | | | |
|---|---|---|---|
| Standard, J. | Oil Stove | Oct. 29, 1889 | 413,689 |

| | | | |
|---|---|---|---|
| Standard, J. | Refrigerator | July 14, 1891 | 455,891 |
| Stewart & Johnson | Metal Bending Machine | Dec. 27, 1887 | 375,512 |
| Stewart, E.W. | Punching Machine | May 3, 1887 | 373,698 |
| Stewart, E.W. | Machine for Forming Vehicle Sear Bars | May 22, 1887 | 373,698 |
| Stewart, E.W. | Station Indicator | June 20, 1893 | 499,895 |
| Sutton, E.H. | Cotton Cultivator | Apr. 7, 1874 | 149,543 |
| Sweeting, J.A. | Device for Rolling Cigarettes | Nov. 30, 1897 | 594,501 |
| Sweeting, J.A. | Combined Knife & Scoop | June 7, 1898 | 605,209 |
| Taylor, B.H. | Rotary Engine | Apr. 23, 1878 | 202,888 |
| Taylor, B.H. | Slide Valve | July 6, 1897 | 585,798 |
| Temple, L. | Toggle Harpoon | 1848 Eyewitness Black History | |
| Thomas, S.E. | Waste Trap | Oct. 16, 1883 | 286,746 |
| Thomas, S.E. | Waste Trap for Basins, Closets, etc. | Oct. 4, 1887 | 371,107 |
| Thomas, S.E. | Casting | July 31, 1888 | 386,941 |
| Thomas, S.E. | Pipe Connection | Oct. 9, 1891 | 390,821 |
| Toliver, G. | Propeller for Vessels | Apr. 28, 1891 | 451,086 |
| Tregoning & Latimer | Globe Supporter for Electric Lamps | Mar. 21, 1882 | 255,212 |

| | | | |
|---|---|---|---|
| Walker, P. | Machine for Cleaning Seed Cotton | Feb. 16, 1897 | 577,153 |
| Walker, P. | Bait Holder | Mar. 8, 1898 | 600,241 |
| Waller, J. N. | Shoemaker's Cabinet or Bench | Feb. 3, 1880 | 224,253 |
| Washington, W. | Corn Husking Machine | Aug. 14, 1883 | 283,173 |
| Watkins, Isaac | Scrubbing Frame | Oct. 7, 1890 | 437,849 |
| Watts, J.R. | Bracket for Miners' Lamp | Oct. 7, 1890 | 437,137 |
| West, E.H. | Weather Shield | Sept. 5, 1899 | 632,385 |
| West, J.W. | Wagon | Oct. 18, 1870 | 108,419 |
| White, D.L. | Extension Steps for Cars | Jan. 12, 1897 | 574,969 |
| White, J.T. | Lemon Squeezer | Dec. 8, 1896 | 572,849 |
| Williams, C. | Canopy Frame | Feb. 2, 1892 | 468,280 |
| Williams, J.P. | Pillow Sham Holder | Oct. 10, 1899 | 634,784 |
| Winn, Frank | Direct Acting Steam Engine | Dec. 4, 1888 | 394,047 |
| Winters, J.R. | Fire Escape Ladder | May 7, 1878 | 203,517 |
| Winters, J.R. | Fire Escape Ladder | Apr. 8, 1879 | 214,224 |
| Wormley, J. | Life Saving Apparatus | May 24, 1881 | 242,091 |

# *Bibliography*

Haber, Louis. *Black Pioneers of Science and Invention.* New York: Harcourt Brace & Company, 1970.

Hayden, Robert C. *9 African-American Inventors.* Maryland: Twenty First Century Books, a Division of Henry Holt & Co., Inc., 1972.

Howell, Ann C., and Grace C. Massey. *Black Science - Communication.* Illinois: Chandler/White Publishing Co., Inc., 1986. Series

Klein, Aaron E. *The Hidden Contributors: Black Scientists and Inventors in America.* New York: Doubleday & Company, Inc., 1971.

McKinley, Burt, Jr. *Black Inventors of America.* Oregon: National Book Company, 1989.

# About the Author

David Maynard Foy is an ordained African-American Minister, in Christian Ministry for twenty years. He attended a church where the main educational thrust was a national agenda to eliminate racism. Its commission on racial and social justice was the instrumentality which attempts to implement change in racial attitudes. He is an active member of this commission.

Before his transfer from the African Methodist Episcopal Zion Church, he led a congregation into a new church facility in Kittrell, North Carolina, an historical community which once educated Blacks at Kittrell College, an institution of the African Methodist Episcopal Denomination.

His educational background includes a degree in sociology from North Carolina A&T State University, Greensboro, North Carolina, a Master of Divinity degree at Shaw Divinity School, Raleigh, North Carolina, and continuing education courses toward a Master of Education degree at North Carolina Central University, Durham, North Carolina.

In 1980, on the campus of Saint Augustine's College, Raleigh, North Carolina, he founded an affiliate chapter of the National Black Child Development Institute.

His present church involvement includes being an Assisting Minister at the Historic Saint Paul A.M.E. Church, founded in 1848 in Raleigh, North Carolina, and an affiliation with the Interfaith Alliance of Washington, DC. This national alliance was formed in 1994 to be a voice opposite the Religious Right Coalition.

Reverend Foy believes that students of various institutions of learning, whether public, private or religious based; will benefit from a broader understanding of the multiple discoveries and inventions introduced to our society by African-Americans. Regardless of a student's ethnicity, a deeper knowledge of so many facts which have been concealed is even vital to the future growth and collective understanding of all who live and function in the American way of life. This project was initiated, continued, and concluded with those beliefs in mind.

Previous works by David Foy include *Human Issues and Human Values*, Council of Concerned African-American Christians, published in 1978.